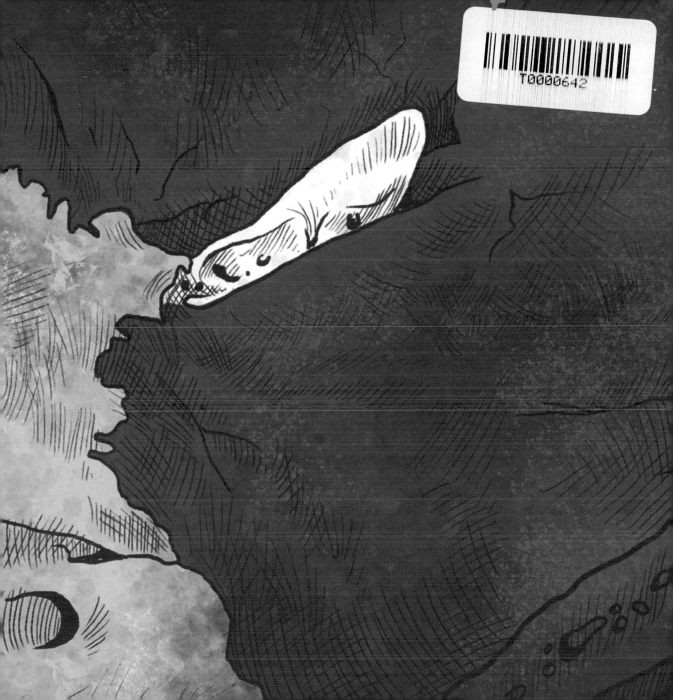

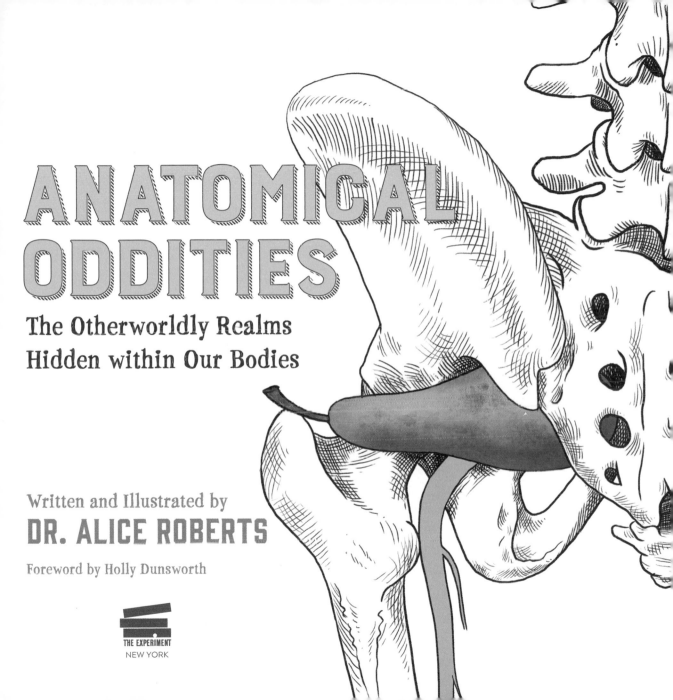

ANATOMICAL ODDITIES

The Otherworldly Realms
Hidden within Our Bodies

Written and Illustrated by
DR. ALICE ROBERTS

Foreword by Holly Dunsworth

THE EXPERIMENT
NEW YORK

The Experiment, LLC
220 East 23rd Street, Suite 600
New York, NY 10010-4658
theexperimentpublishing.com

THE EXPERIMENT and its colophon are registered trademarks of The Experiment, LLC. Many of the designations used by manufacturers and sellers to distinguish their products are claimed as trademarks. Where those designations appear in this book and The Experiment was aware of a trademark claim, the designations have been capitalized.

The Experiment's books are available at special discounts when purchased in bulk for premiums and sales promotions as well as for fundraising or educational use. For details, contact us at info@theexperimentpublishing.com.

Library of Congress Cataloging-in-Publication Data

Names: Roberts, Alice, 1973- author.
Title: Anatomical oddities : the otherworldly realms hidden within our
    bodies / written and illustrated by Dr. Alice Roberts.
Description: New York : The Experiment, [2023] | Includes index.
Identifiers: LCCN 2023021501 (print) | LCCN 2023021502 (ebook) | ISBN
    9781891011139 | ISBN 9781891011146 (ebook)
Subjects: LCSH: Human anatomy--Popular works. | Human anatomy--Pictorial
    works. | Human body--Pictorial works. | Human body--Popular works. |
    Human physiology--Popular works.
Classification: LCC QM26 .R63 2023 (print) | LCC QM26 (ebook) | DDC
    612--dc23/eng/20230601
LC record available at https://lccn.loc.gov/2023021501
LC ebook record available at https://lccn.loc.gov/2023021502

ISBN 978-1-89101-113-9
Ebook ISBN 978-1-89101-114-6

Cover and text design by Beth Bugler

Manufactured in China

First printing November 2023
10 9 8 7 6 5 4 3 2 1

For Jonathan

ἀρχὴ παιδεύσεως ἡ τῶν ὀνομάτων ἐπίσκεψις

"The beginning of education is the investigation of terms."

Antisthenes (446–366 BCE)

CONTENTS

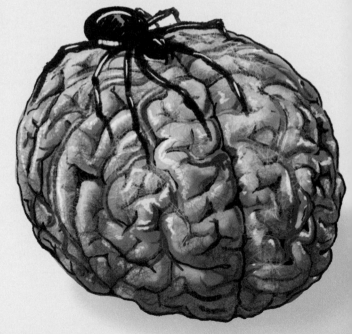

FOREWORD BY HOLLY DUNSWORTH

My family's idea of a fun Friday night is enjoying a nice meal around the kitchen table and then googling "teratoma" together, reveling in the images of tumorous balls of hairy teeth. But there is appreciating the patently strange—like the teratoma, which comes from the Greek for "monster"—and then there is turning the familiar strange. By investigating the names of more typical tissues, Alice Roberts makes everything from arbor vitae cerebelli (our brain's tree-like white matter; page 37) to zygoma (a bone we all have in our faces; page 109) just as surreal as monstrous tumors, but even more astonishing—in part because closer examination yields fascinating detail, and in part because these body parts, common to us all, cooperate to keep us alive right now and into the future, and have done so for millions of years.

To name anatomy is to tilt at understanding. The language of our bodies comes from ancient people who lugged olive oil to bathhouses, who made flutes out of leg bones, and who scrawled with metal spikes on wax tablets. Hence, brains have no "motherboards" or "microchips." But anatomy's originators did stick mice under our skin (page 54), a cock's comb on our head (page 101), and a horse's tail near our derriere (page 46). Ravens are perched within each shoulder, keeping watch over our spider hands (pages 50 and 81). Anatomists are Frankenstein on psychedelics.

As a student writing about fossil ape tarsals, which are foot bones, I saw them everywhere. Pebbles on a gravel path and in a pond. Potatoes on my plate. Animal droppings. When I described one "like a submarine," my supervisor scribbled "the boat or the sandwich?" This book makes clear that anatomists live in a world of visual metaphor. For them, a hip muscle isn't just a hip muscle when it so strongly resembles a pear—hence its name, "piriformis," for pear-shaped. I wonder, for anatomists, after so many dissections of the muscle, when does a pear become more muscle than fruit? Do anatomists hiking among Earth's lungs see only bronchial trees (page 26)?

O! How an art-full, word-mad book about the body can exercise the mind. The most mundane moments in life, suddenly interrupted and elevated. Just brushing my teeth, I see the bathroom mirror is now enchanted. Inspecting my soon-to-be-gray temples, I slow my scrubbing to a hush and ponder how "to stretch" (*temp-*) and "time" (*tempus*)—both reverberating in the muscle "temporalis" and the "temporal" bone (page 98)—are really just one and the same. How transcendental.

HOLLY DUNSWORTH is a biological anthropologist at the University of Rhode Island, where her research and teaching focuses on scientific narratives of human evolutionary history.

BEFORE

WE

START...

The human body has been ogled, scrutinized, surveyed, chopped up and put back together, chopped up and not put back together; parts of it have been cut off, sliced up, put under the microscope; zapped with X-rays, buzzed with ultrasound, pierced with rods containing lights, tools, and cameras; mummified, dried, plastinated, pickled. In the process, people have seen it from new angles, in new detail. And when you explore a landscape—and discover a new feature in it—what do you do? You name it.

The human body is littered with names so densely, not an inch of it escapes attention. Some of these names reflect what things look like—the hippocampus of the brain looks a bit like a tiny seahorse; the mammillary processes of the vertebrae look like miniature boobs. Others tell you what things do—levator labii superioris lifts your upper lip; levator ani lifts your butthole. Some anatomical features are named after their discoverers—or at least, the people who first wrote about them in detail. And many of these sound like mysterious places you might want to visit—the islets of Langerhans—or not—the crypts of Lieberkühn.

I've anatomized innumerable bodies and I've always drawn anatomy—to learn it myself, and to teach it to medical students. And sometimes I would make quirky drawings just to help me remember the names of things—I've always preferred visual to verbal mnemonics. In this book, I've set out to explore the human body as a miniature explorer, a Lilliputian embarking on a voyage, to retrace stories of discovery in human anatomy and to draw the landmarks that exist in the landscape of the body. Shrink yourself to a minute scale and the villi of the intestine look like looming, grassy mounds, while the islets of the pancreas look like stromatolites on a shore and the hollow space inside the upper jawbone looks like a dark, foreboding cavern. Some of my illustrations are more about the meanings of and inspirations for anatomical terms, or are just a bit weird—because anatomy is weird, and wonderful.

I've also tracked down the origins of terms, bringing out hidden messages and cultural allusions in the names, imagining the torn tendon of Achilles in an image on a black-figure vase, the Atlas vertebra holding up the world, and the duodenum as an alien creature with twelve fingers. Around 10 percent of words in spoken English come from ancient Greek, while half are Latin, and the rest are Old English. There's no hard and fast divide between these languages, though, as many Greek words entered Latin in antiquity and Germanic languages come from the same original source as the Romance languages—a very ancient, Bronze Age tongue known as Proto-Indo-European. (We have no direct evidence of Proto-Indo-European, or PIE, but linguists have reconstructed sounds and whole words based on similarities between spoken and historically attested Indo-European languages. The reconstructed words are always marked by a preceding asterisk.) Through time, there would have been many words shared, stolen, and borrowed between different languages. Even so, medical terminology is quite different from colloquial English. It is much more Greek—as this was the scholarly language of antiquity; around two thirds of words in medical terminology are essentially Greek passed through a Latin filter—sometimes via Arabic. This means that some words have become distracted from their original meaning, lost in translation. (The ancient Greek anatomist Herophilus almost certainly never thought that the large veins inside the skull bore any resemblance to a wine press, as we shall discover). But it also means that we might learn something important about a certain anatomical structure, and what its discoverers or namers thought of it, if we can dig back into that history of words, into their etymologies.)

The word "etymology" itself has an interesting etymology. It comes from Greek: *logos*, meaning "study," and *etymon*, meaning "truth." It's been used in the sense of "true word" since at least the sixteenth century—and carries with it an idea that the origin of a word may reveal its true meaning. Of course, words and their meanings change through time, and with scientific terminology in particular, we are always primarily concerned with precise and current definitions when we write academic papers and medical reports. But the etymology of scientific terms does reveal a kind of truth—it shows us what the early anatomists were focusing on when they thought up these names. It can make us think and look again at a particular piece of anatomy. It can also help us to remember the words.

CRYPTS OF LIEBERKÜHN

I wanted to see my insides. I swallowed a large pill, gulping it down with water, and waited for the results. It was a PillCam, and it provided me with a strangely intimate and revealing experience, beaming pictures of my intestines out of the dark recesses of my body onto a glowing screen. The PillCam kept track of its path, too, and so I learned that my duodenum loops in a different way from most people's—an anatomical oddity that I had no idea about, and which doesn't signify anything beyond the fact that we're all a little bit different, outside and in. There's no single ideal, no peak of "normal," just a spread of variation. Even if your duodenum doesn't twist like mine, the lining of your intestine would look similar, if you were to swallow a camera to see it. Lit by a tiny flash, the PillCam revealed the pink lining of my duodenum, jejunum, and ileum, and although the resolution wasn't all that high, there was the impression of *fluffiness*, like velvet or perhaps a shag-pile carpet. And that's because the surface is entirely covered in minute filaments. They're called villi, from the Latin for "shaggy hair," which is very apt.

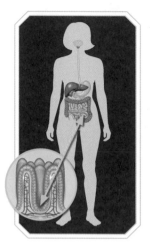

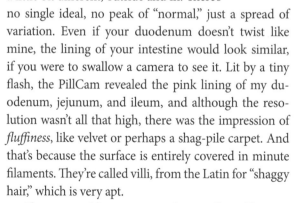

If you were able to zoom in closer still, you'd see an epic landscape—those shaggy filaments looking more like tall mounds, and each of *these* being shaggy, too. The villi are covered with microvilli. And down around their bases you'd see the dark openings of deep pits: well-like intestinal crypts. Everything is here for a reason. The shaggy villi with their shaggy microvilli increase the surface area of the intestine—important for absorbing as much as you can from the digested contents of your guts. The deep wells are lined with cells which busy themselves making a cocktail of hormones, chemical signals to control digestion, and antibacterial fluid to eliminate unfriendly bacteria that happen to have wandered into you this far. The word "crypt" is perhaps more familiar from churches or tombs, where it describes an underground chamber. It comes from Greek, via Latin—*crypta*—and its original meaning is "hidden" or "secret." An eighteenth-century German physician by the name of Johann Nathanael Lieberkühn was the first to uncover these secrets and write about them in detail, in his rip-roaring book on the "fabric and functions of the villi of the human small intestine." He studied the microscopic structure of various body parts, inventing one microscope that could shine light onto thick, opaque tissues, and another to watch blood flowing through vessels. His contemporaries called these contrivances *Wundergläser*—"wonder glasses." And the intestinal crypts are still sometimes referred to as the crypts of Lieberkühn.

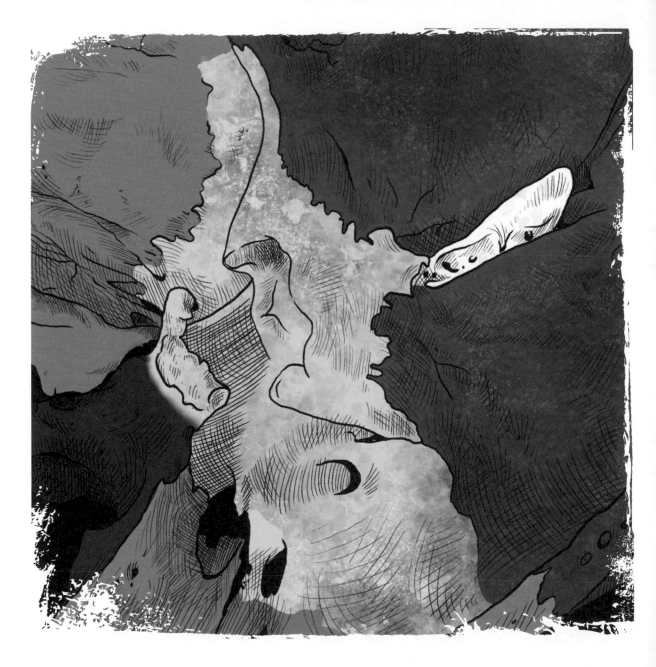

SELLA TURCICA

The pituitary gland is only the size of a pea, and yet it's extremely important. Its name means "secreting mucus," and this seems to be because it was originally thought to be the source of snot—it lies inside the skull, just above and behind the upper recesses of the nasal cavity. But in fact that's not what it does at all. It's a small but crucial part of the body's endocrine system of chemical messaging. It pumps out a variety of hormones into the bloodstream, which mainly go off and tell endocrine cells elsewhere in the body—in the thyroid and adrenal glands, the ovaries, or the testes—to get on with pumping out their own hormones.

Such an important gland deserves a suitable throne to sit on, and that's what the sella turcica is. The pituitary gland hangs down under the brain and sits comfortably on the top of the sphenoid bone, right in the center of the base of the skull, in a small depression. This cavity is named after a Turkish saddle, because it curves

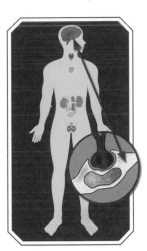

up at the front and the back, just like the pommel and cantle of its namesake, holding the rider securely on the horse's back. In the time of the Ottoman Empire, a horse with a beautifully decorated Turkish saddle—possibly embellished with gold or silver—would be the best gift you could ever hope for. But everyone has a miniature version—tucked away inside your skull, just under your brain.

There it is—in the center of this image, right in the middle of the sphenoid bone, picked out in lime green.

ACROMION

This prong of bone juts out over the shoulder. Just as the Acropolis is the "upper city" towering over the rest of Athens, the acromion is the highest point of the shoulder blade—from the Greek *akros*, meaning "high," and *ōmos*, meaning "shoulder." We first learn this term from the compilation of medical texts, from the fourth century BCE, attributed to Hippocrates (but actually representing a collective body of work). An old word for the *scapula* (Latin for "shoulder blade") was the Greek-derived "omoplate," with *plate* meaning "flat" or "broad."

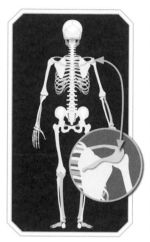

The acromion is a continuation of the spine of the scapula, which you can feel quite easily if you place one hand up and over the opposite shoulder. A sharp bar of bone runs diagonally, and as you trace it laterally, or outward, it ends in a broader wedge—the acromion. You should also be able to feel the joint where the front of the acromion articulates with the lateral end of your clavicle, or collar bone (clavicle itself coming from the Latin for "little key"). The acromion isn't officially part of the shoulder *joint*—the glenohumeral joint—but it offers a stabilizing presence to it.

The shoulder joint is the most mobile joint in the body—and therefore the most prone to dislocation. When it dislocates, the head of the humerus or upper arm bone moves in a downward direction; the acromion stops it displacing upward. The acromion also provides an opportunity for muscle attachment; the middle fibers of the triangular muscle deltoid (from the Greek for "delta-shaped") are anchored on it. But the acromion can be a pain as well. Between the undersurface of the acromion and the top of the humeral head, there's a fairly narrow gap, which accommodates the tendon of the long head of biceps muscle, the shoulder capsule itself, the fleshy tendon of supraspinatus muscle, and a little pocket of lubricating fluid called the subacromial bursa. The narrowness of the subacromial space means that the structures occupying it can be subject to impingement. The bursa may become inflamed. Supraspinatus and the biceps tendon may become frayed and even tear—problems like these are often lumped together under the term "rotator cuff syndrome," which is very common. I did a PhD on just this pattern of pathology, looking at evidence for it in ancient bones and other apes, and somewhat predictably, have noticed my own shoulders starting to show signs of this degenerative disease. If I keep my muscles in good shape, it's not too troublesome—and that of course applies to any joint, as we get older. Use it or lose it, my friends. Now stop reading and do ten push-ups!

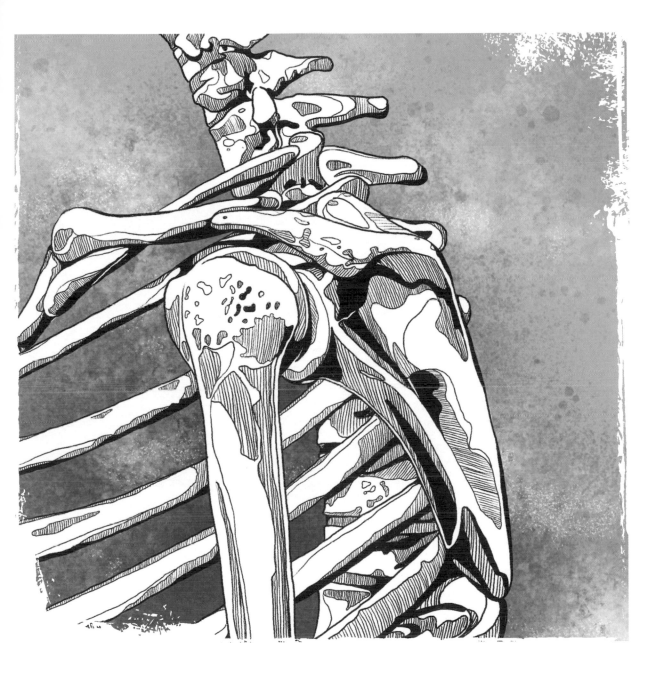

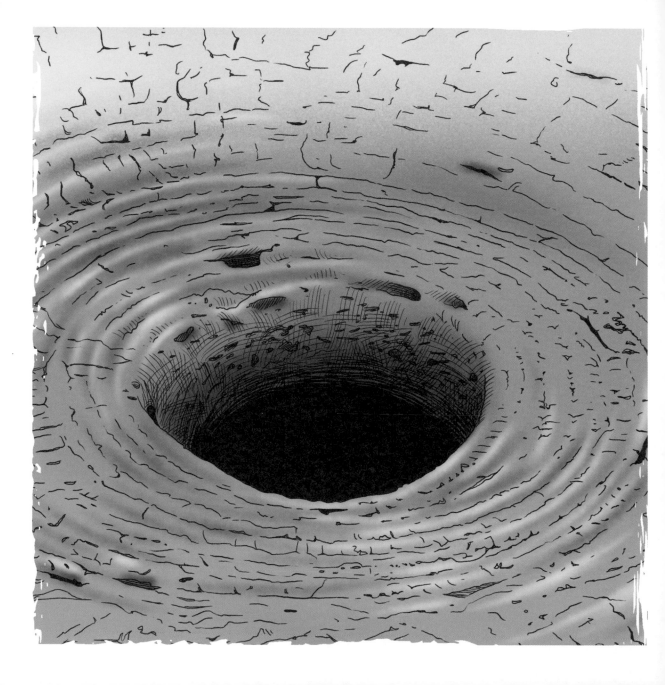

HAVERSIAN CANAL

It's easy to think of bones as dead, white structures. Even if you've never seen a real human skeleton, you will undoubtedly have seen skeletons in photographs, on television, and in films. And you'll be aware that bones tend to stick around in the ground—longer than any other body part (except teeth)—while everything else decays away. Because of this, we can end up thinking of bones as chalky, inert objects—rather like seashells. But living bone is quite different. Firstly, the skeleton is part of a wider system of body tissues—bones are linked together by cartilage, membranes, and ligaments, and can't be understood without knowing about the muscles that attach to them. Secondly, living bone is not white and inert; it is pink and dynamic. It is full of cells that are constantly monitoring stresses and strains, and then adding more bone mineral where it's needed, etching it away where it isn't. It's pink because it's full of blood vessels. If you've ever broken a bone, you'll know this very well. The torn vessels will gush blood if the bone breaks through the skin, or create a large swelling as blood pours into surrounding tissues. Within the shafts of long bones like the humerus and femur, networks of blood vessels run through microscopic, longitudinal channels called Haversian canals.

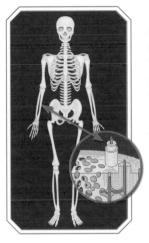

This image is an up-close view of a single Haversian canal, surrounded by concentric layers, or lamellae, of bone mineral. The canals are named after Clopton Havers, a British doctor who published a book on bones in 1691: *Osteologia Nova*. In it, he described the detailed features he'd observed with the help of a magnifying glass, including minute "longitudinal pores." (He didn't see any blood vessels in them, though, and thought that they existed to help oil from the marrow permeate into the bone. Never mind, at least he *saw* them—and he would have been utterly astonished at the detail we can see today, with our high-powered microscopes.)

The small cavities you can see here between the lamellae are lacunae (Latin for "pit" or "pond"), each inhabited by a single bone cell, or osteocyte. Those osteocytes seem lonely, trapped inside their caves, surrounded by hard bone mineral. But they reach out to their neighbors via tiny, hair-like projections, through minute mini-canals—canaliculi—making contact with other osteocytes. In this way, they communicate with each other, and with other cells, which respond by laying down more bone mineral or taking it away. That conversation between bone cells means that if you become less active, you will lose bone mass, but if you exercise more, your bones will respond by becoming stronger. Yet another reason for staying fit.

SPHINCTER

A sphincter says 'what'?" goes the joke. It's too easy to be tripped up by this particular pitfall, to fall flat on your face in the "Oh!" of the "O." It's funny—of course—because the sphincter we all think of before any other is the one in your butt, the anal sphincter. And there's nothing as enduringly humorous as the nether regions, when it comes to anatomical amusements. You may laugh, but do you even know what

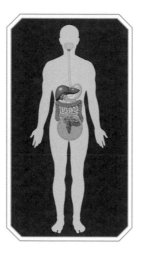

a sphincter is? Well, if you don't, you're about to find out. It's a doughnut-shaped ring of muscle that can relax or tighten to open or close the hole in its center. The word comes from the Greek for a constricting band, from *sphingein*, "to bind tight" (and this must surely be related to the Greek word for wasp, *sphēx*—with its cinched-in waist).

There are many, many sphincters in your body, not just the anal one that opens to let feces out and closes to keep it in. There are several others along the tube of your gut, including the pharyngeal sphincter, or upper esophageal sphincter, which opens to allow food into your esophagus when you swallow, but stays closed the rest of the time to stop you swallowing air all the time. The lower esophageal sphincter lies at the entrance to the stomach; the pyloric sphincter at the exit from the stomach, opening to let churned-up food into the duodenum; the ileocecal valve or sphincter at the junction between the small and the large intestines.

There are also sphincters in your blood vessels—little circles of muscle that guard the entrance to networks of capillaries and can be closed off to regulate blood flow. There's a urethral sphincter controlling the flow of urine out of the bladder into the urethra. While most sphincters open and close to control the flow of fluid, there's one that works with light. It's the sphincter pupillae—the smooth muscle of the eye's iris that constricts your pupils.

Back in the guts, there is the oddest sphincter of all: the sphincter of Oddi. It lies at the base of the bile duct, opening to release bile from the gallbladder into the duodenum when we've eaten a meal. Bile helps to break down oils in the foods you've eaten. A twenty-three-year-old, Ruggero Oddi, first described this sphincter, back in 1887.

sphincter

TENDO ACHILLES

What has a tendon at your heel got to do with one of the heroes of the Trojan War? Why is it called the Achilles tendon? It wasn't discovered by Achilles, certainly—so how did the name come about?

Achilles, so the myth goes, had a dodgy ankle. Or at least, a particularly vulnerable ankle. His mom, a sea nymph named Thetis, tried to make him invincible by dipping him, as a baby, in the River Styx. Surprisingly, this seems to have worked very well—apart from the fact she'd held him by an ankle to dunk him, and so this small part of him escaped being magically shielded. Years later, fighting with the Greeks, Achilles suffered a fatal wound. There are many different versions of this tale, with different details. Only the later ones got the memo about Thetis and the Styx, and focus in on his heel being the crucial target of a Trojan arrow.

But it would be three millennia before the thick tendon that attaches the calf

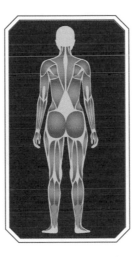

muscles to the heel bone, or calcaneum, would become linked with the legendary Achilles. In 1693, a Dutch anatomist, Philip Verheyden, carried out the rather gruesome operation of dissecting his own amputated leg. And writing about it. Waxing lyrical, he described the calcaneal tendon as the "chorda Achillis," or Achilles' sinew. In the early eighteenth century, a German professor of anatomy called Lorenz Heister called it "tendo Achillis," the Achilles tendon. (His own name lives on in the spiral valves of Heister, in the duct of the gallbladder.)

The vulnerability implied by the name is real. The tendon is the thickest in the body, but also one of the most commonly ruptured, especially in runners. This shouldn't put you off running, but it's a good idea to do some stretching, do some lower impact sports as well, and build up to running longer distances gradually. Or else you might end up like this flayed Achilles depicted—just as his tendon snaps—on a black-figure vase.

This is a hole in the head, quite literally. Foramen means "hole" in Latin, from the verb *forare*, "to bore." In the way that language has this habit of evolving, unfurling in different directions and then reuniting, *forare* and "bore" really are the same word—or at least, linguistic cousins, coming down to us through different routes but rooted in the same ancient, Proto-Indo-European word: **bhorh*.

There are many foramina in the head. The base of the skull is peppered with them, allowing cranial nerves to escape from the bony confines of the cranium and reach their destinations—in muscles, skin, or glands. But there are also small foramina in the brain itself, allowing the fluid that is made in its ventricles (see page 102) to escape. Your brain makes about two cups (half a liter) of cerebrospinal fluid, or CSF, each day, and it has to go somewhere. In the lowest part of the brain, there are three holes: the foramen of Magendie in the midline, and a foramen of Luschka to each side. CSF flows out through them into the space around the brain and spinal cord and is eventually reabsorbed back into large veins lying within the meninges, around the brain. The brain effectively floats in a bag of CSF—so it doesn't crush the nerves and vessels lying underneath it. CSF also helps to cushion the brain itself, protecting it from

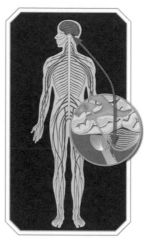

impacts, as well as removing waste products from brain metabolism. The brain is constantly leaking its waste into its own swimming pool, if you like, but that pool is continuously refreshed, so it's not as disgusting as it sounds. But if the exits from the ventricles—the foramina of Magendie and Luschka—are blocked, then the pressure of CSF builds inside the brain, with potentially disastrous consequences, as delicate neurons become compressed.

The midline foramen, opening just below the cerebellum, bears the name of a pioneering French physiologist who lived around the turn of the nineteenth century: François Magendie. He made many discoveries, but his thirst for new knowledge led him to carry out some horrific experiments on animals. He killed one dog by feeding it just sugar for a month, and discovered the function of the cranial nerves by undertaking vivisection on various other dogs. Knowledge can be hard-won, but his techniques attracted criticism even at the time. In Balzac's 1831 novel *La Peau de Chagrin*, a thinly disguised version of Magendie appears as Docteur Maugredie—"Doctor 'Bad Grace'"—a man of "distinguished intellect, but skeptical and contemptuous." The landscape of anatomy was mapped through centuries of detailed observation and analysis but is also liberally pierced with tales of pain and suffering.

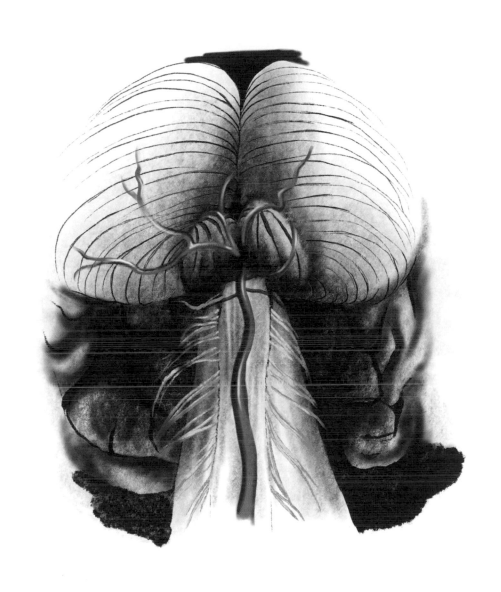

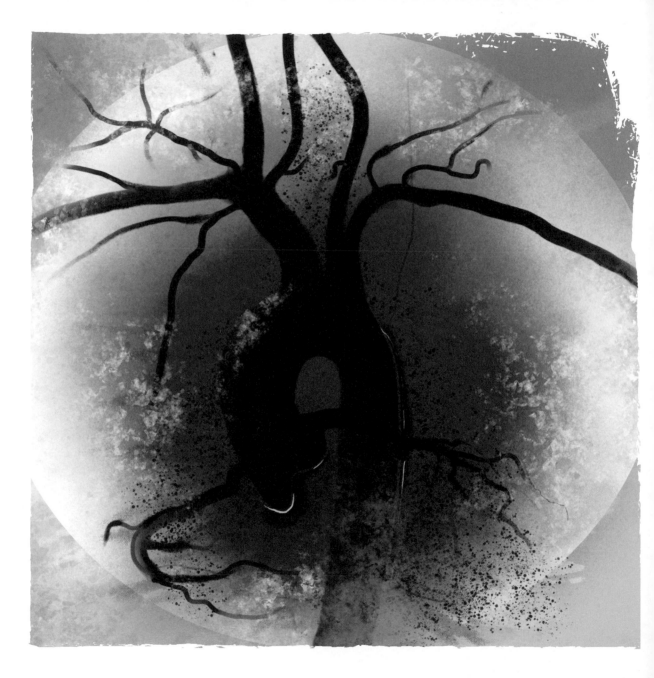

AORTIC ARCH

The aorta, the largest artery in the entire body, leaves the heart as the ascending aorta, then proceeds to do a U-turn, forming the arch of the aorta, turning downward as the descending aorta, which runs down just in front of your spine, all the way to the level of the first lumbar vertebra, where it ends by dividing into two vessels, the common iliac arteries.

It's an odd word, aorta. It comes from Greek, *aorte*, meaning "to raise" or "hang up." The word *aorter* refers to the shoulder strap used by Greek soldiers to suspend their swords. Perhaps if you dissected a heart out of a chest together with the aorta, you could use the vessel to hang up the heart? Who knows. But there's also a close connection with the word "artery," which comes from the Greek *arteria*. And this word introduces another element— and a case of mistaken identity—into the story. The word *arteria* is made up of *aer* and *terein*—it means the "air keeper." And that's because an ancient theory about arteries had it that they carried air or *pneuma*, not blood at all. It's one of those bizarre theories that seems extremely easy to disprove—

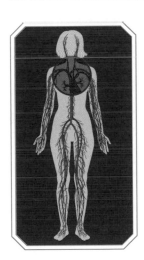

cut into an artery and you find out soon enough what it contains—but it hung around for a very long time. Just to confuse things even more, Hippocrates was the first to use the word *aorta*— but he used it for the trachea, or windpipe, and the bronchi, which really are full of air. We can blame Aristotle for introducing the muddle— transferring the word across to the great vessel, where it stuck. Then we have centuries of both the trachea and the aorta being called artery. The windpipe was distinguished as the "rough artery"—the *arteria tracheia*.

So the aorta isn't full of air and it doesn't really hold the heart up. But it does have an arch.

BRONCHIAL TREE

The lungs bud off from the gut of a human embryo, like a shoot budding off a root. Branching from the hollow tube of the primordial gut, the bud itself is hollow. It soon branches itself, into two lung buds. Each bud continues branching and branching, growing into a tree of tiny connected tubes. These will form the bronchi and smaller bronchioles of the lungs. Eventually, in the last two months of gestation, the tips of the tubes form clusters of minute, bubble-like alveoli, wrapped in capillaries. This is where gas exchange will happen—oxygen from inhaled air will pass into blood, carbon dioxide will move out—as soon as the baby is born and takes its first breath. Inside the womb, though, the lungs are full of amniotic fluid, which is drawn in and pushed out by contractions of the diaphragm, as the fetus practices the movements of breathing.

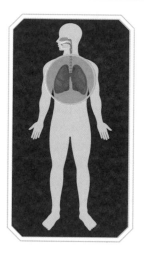

The word "bronchus" comes from the Greek *bronkhos*, which originally meant "windpipe." Now, we differentiate between the trachea and its branches—the bronchi. But *bronkhos* is also related to a word meaning "wet," or "moist," with deeper roots, coming from a Proto-Indo-European word meaning "to devour." So, a historical misconception is trapped in the word "bronchus," and there's a symmetry of misunderstanding: Arteries were once believed to convey air, while the airways were once thought to contain fluid. The "Father of Medicine," Hippocrates (460—370 BCE), instructed students at his school on the island of Cos to listen to their patients' breathing. And yet it was thought that air—*pneuma*—flowed in through the nostrils and then to the brain and through vessels to the rest of the body. In the fourth century BCE, Herophilus led the school of anatomy at the famous University of Alexandria in Egypt. He undertook human dissection, discovering that air was drawn through the trachea and bronchi into the lungs—although he still believed that it was then distributed via arteries. The Roman physician Galen saw what others had not: that arteries contained blood, not air. By the sixteenth century, some clarity emerged—the pulmonary and systemic circulations were well mapped and it was clear that *blood* flowed through the lung's capillaries while air was drawn down into bronchi, bronchioles, and alveoli—but no one knew *why*. In the seventeenth century, a Cornish doctor, John Mayow, worked it out. Writing in 1674, a hundred years before the discovery and naming of oxygen, Mayow concluded that the chief function of breathing was "that particles of a certain kind, absolutely necessary for the support of animal life, may be separated from the air by means of the lungs and mixed most minutely with the mass of the blood."

DUODENUM

The first part of the small intestine, just after the stomach, is called the duodenum. It is about 10 inches (25 cm) long and forms a C-shaped loop, running right, then down, then up and left, before turning forward to become the second part of the small intestine, the jejunum. Churned-up food from the stomach, a thick, sludgy goop known as chyme (from Greek, *khumos*, "juice"), oozes out through the pyloric sphincter into the duodenum, ready for further digestion. That sphincter is named from the Greek, *pulouros*, for "gatekeeper." Once the duodenum fills with chyme, it sends chemical messengers via the bloodstream to the liver, gallbladder, and pancreas to stimulate them to push bile and digestive enzymes down their respective ducts, into the duodenum. One of these chemical messengers is known as secretin—because it stimulates secretion—and it was the first hormone to be discovered, back in 1902, by English physiologists William Bayliss and Ernest Starling.

Having had those digestive juices added to it, the now-watery chyme passes along the rest of the small intestine, from the duodenum into the jejunum and then the ileum, with all the broken-down

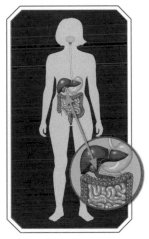

nutrients being absorbed into the bloodstream. Eventually, the chyme enters the large intestine, where a bit more digestion of starch goes on, and water is absorbed to make the gut contents progressively more solid, ready for its exit from the body. The word "jejunum" comes from Latin for "fasting" or "empty"—apparently because it is usually found empty at death (though perhaps this is just because its contents are less solid than the chyme in the stomach or the large intestine). The word "ileum" comes via Latin from Greek, *eileos*—which also refers to gripping colic. The roots of that Greek word mean "to turn" or "revolve," which aptly describes the looping small intestine, but also the twisting and squeezing pain of a colicky bellyache.

What about "duodenum"? This word passes through monstrous contortions to reach us. It comes from Latin *duodeni*, meaning "twelve each"—from *duodecim*, "twelve" (which also gives us "dozen"). But twelve what? Fingers, it turns out—*duodenum digitorum*. And for this we can blame Herophilus, who called it the *dodeka-daktylon* in Greek—the twelve-fingers-long part of the gut. But who has twelve fingers to measure it by?

ISLETS OF LANGERHANS

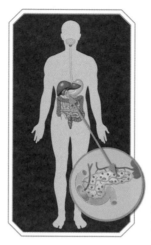

Here are the pancreatic islets as you've never seen them before (if you've even ever seen them before). They look like those weird stromatolites on the coast of western Australia, mini-reefs built up over centuries by mats of microbes. But these are made of cells—very special cells in the pancreas that dedicate their lives to producing the hormone insulin. While the cells around them are busy making the enzymes that will run down the pancreatic duct and into the duodenum to break down the proteins, starch, and fats in chyme, these cells release their product into the bloodstream.

Looking around for a term to describe blood-borne chemical messengers like secretin, Ernest Starling first used the term "hormone" in a lecture at the Royal Society in 1905, where he described some of the pioneering work he'd carried out with his brother-in-law, William Bayliss. He'd looked to Greek for a source of inspiration, and hit upon the word *hormōn*, meaning "impelling" or "setting in motion."

In the 1920s, Canadian scientists managed to extract and purify insulin—from the pancreatic islets of dogs—using it to treat diabetics for the first time. The islets themselves had first been noticed, through the lens of his microscope, by a German medical student, Paul Langerhans. He published his findings in 1869—though he'd had no idea about the function of the "little heaps of cells" he'd seen. After serving as a medic in the Franco-Prussian war, Langerhans became a professor at the University of Freiburg. But then he developed renal tuberculosis. He took leave and traveled to Madeira, where he felt better enough to do some research on marine worms—and even to work as a doctor again, treating many fellow TB sufferers who had also sought out the warm breezes of the island. He died in 1880, aged forty, his legacy remembered in the names of several marine worms, in immune cells in the skin which he'd also spotted as a keen-eyed medical student—and in those islets in the pancreas.

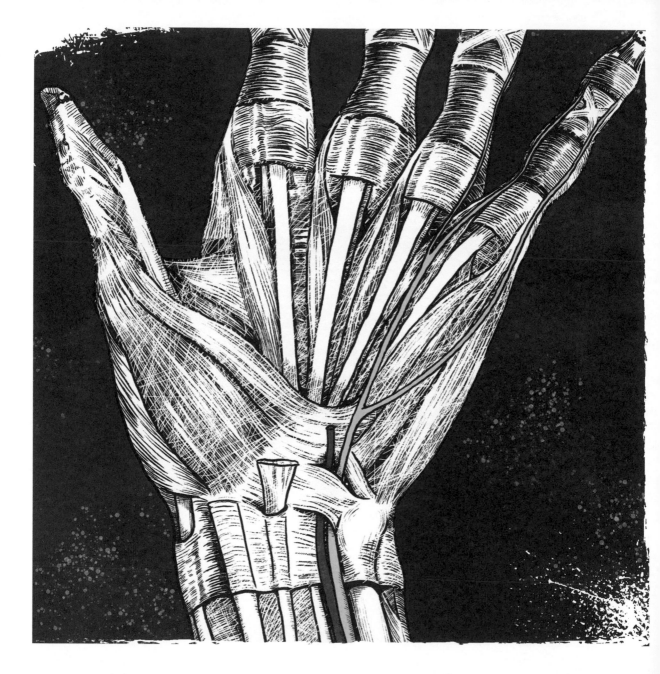

GUYON'S CANAL

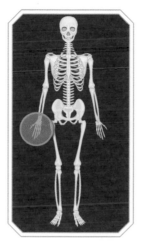

A nother canal. And again, not a watery one. Just a long hole, perhaps more tunnel-like than canal-like in the end. (In fact, it is sometimes known as Guyon's tunnel, or the ulnar tunnel). You should be able to locate it in your own hand: It lies just beyond the wrist, in the "heel" of your hand, on the inner side. Push in with your other thumb and you'll feel a small knobbly bit of bone, the size of a large pea (the pisiform bone—see page 81). Guyon's canal lies just to the thumb-side of that knobble.

To be more anatomically precise, the canal lies distal to the radiocarpal joint, and lateral or radial to the pisiform bone and the hook of hamate. Its roof is the superficial palmar carpal ligament, and its floor is the flexor retinaculum and the origins of the hypothenar muscles. That needs some unpacking if you're unfamiliar with all these terms. "Medial" and "lateral" are extremely useful anatomical terms. Stand up straight, with your arms hanging by your sides and palms facing forward: This is the "anatomical position." "Medial" and "lateral" are relative terms and describe whether something lies closer to the midline of the body (medial), or to the outer edge (lateral). Sometimes it's easier—with hands and arms—to use "ulnar" instead of "medial," and "radial" instead of "lateral," using terms relating to

relative positions of the forearm bones. There are eight small bones in your wrist, or carpus (from the Greek, *karpos*, for "wrist"), and the pisiform and hamate are two of them. More on the pisiform bone later, but "hamate" comes from the Latin *hamatus*, and it does indeed have a little hook sticking out of it at the front (anteriorly, to be more anatomical).

The pisiform bone and the hamate form the medial attachment of a thick ligament that holds the long flexor tendons in place. You can see these tendons clearly at the wrist when you wiggle your fingers around. This ligament is called the flexor retinaculum—"retainer of the flexors"—and stops those tendons bowstringing out when you bend or flex your wrist or fingers. An important nerve supplying the hand, the median nerve, runs under the retinaculum—and can be compressed there, causing pain and tingling in the hand. This is known as carpal tunnel syndrome. Another nerve, the ulnar nerve, runs over the retinaculum, through Guyon's canal. Compression here is rarer, but can still happen—resulting in ulnar-nerve entrapment or Guyon's canal syndrome. Finally, the eponymous name of the ulnar canal remembers the French surgeon Jean Casimir Félix Guyon, who specialized in urological surgery—but clearly knew a thing or two about the hand as well.

THYROID

This is a digital painting of the famous shield frieze at the Palace of Knossos on Crete, based on the reconstruction by archaeological artist Émile Gilliéron. Only very small fragments of the original, Bronze Age fresco survived, with the whole design being reimagined by Gilliéron. There's a lot of artistic license at work, and it's even been suggested that Gilliéron faked some archaeological artifacts, but this type of shield, at least, is not a work of fiction—as we see it depicted authentically elsewhere, on vases and in sculptures. Its figure-eight shape is, I think, what the Roman anatomist Galen had in mind when he wrote that the largest cartilage in the larynx was *thyreoeides*—"shield-shaped." The term also became used for the gland that sits below the larynx, in front of (anterior to) the trachea: the thyroid gland. Both the thyroid cartilage and thyroid gland have a form that involves a pair of lobes connected by a slender isthmus in the midline—like an 8 lying on its side. The word *thyreos* means a "door-shaped" shield—coming from a hypothetical Proto-Indo-European word *dhwer, which becomes *duvara* in Old

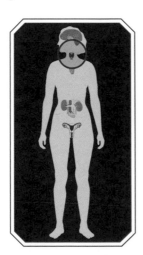

Persian and the more familiar *dor* in Old English. The leading consonant changes in Latin to give us *forum*—for a public space, *outdoors*. The original form of the word is often plural, or double—did those ancient Proto-Indo-Europeans have two-part, stable doors, I wonder?

As for the connecting portion between the two parts of the thyroid gland, "isthmus" is a term we're familiar with from geography, too—where a narrow strip of land connects an island to the mainland. It comes from the Greek *isthmos*, meaning "neck," but the original meaning of the term seems to be "to go across"; it is a connection.

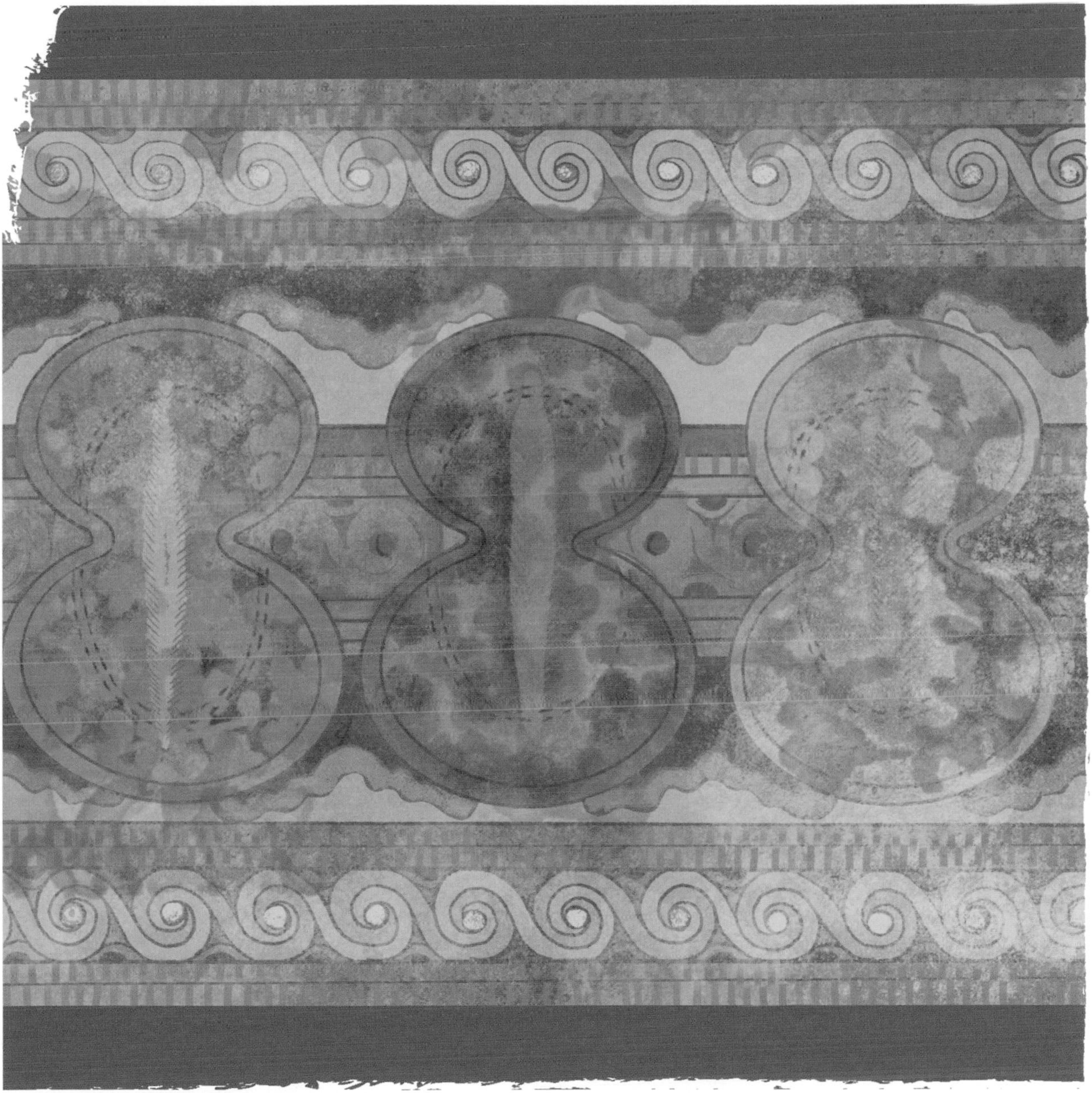

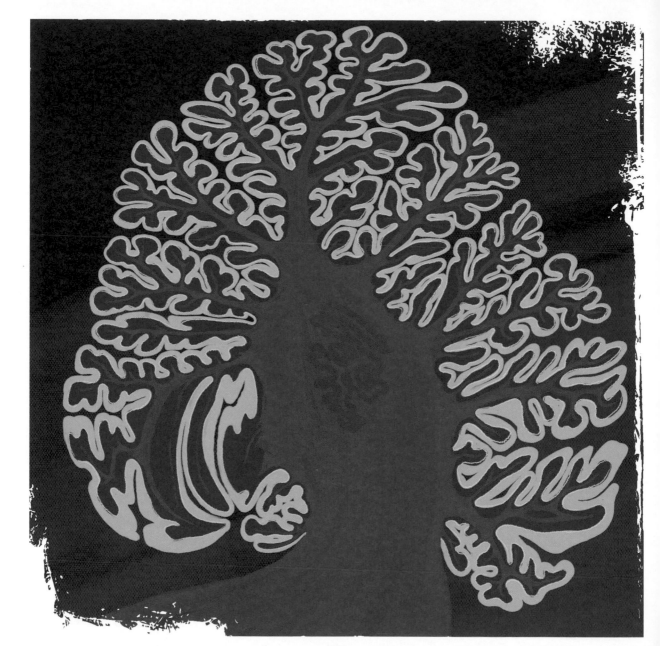

ARBOR VITAE CEREBELLI

When you slice a brain in half, down the midline—something I have done many times—you don't have too much to cut through, as the cerebrum is organized into two hemispheres connected by a narrower bit of nervous tissue—very much like a walnut. But the smaller cerebellum—the "little brain"—which sits underneath the occipital lobes of the cerebrum, at the back, is only slightly indented in the middle. The brain knife—sharp as flint—slices through a thick, bun-like mass, and as it falls into two halves, you see an astonishing pattern inside it: The white matter forms a tree with trunk, branches, and twigs, with the gray matter arranged around that structure like foliage. It's known as the arbor vitae.

Outside anatomy, "arborvitae" is also the name used for an evergreen tree from North America, also known as a white cedar (though no relation of true cedars). When French explorer Jacques Cartier and his crew found themselves sick with scurvy in Canada, in 1536, indigenous people offered them life-saving tea made from this conifer. A frond of arborvitae looks very much like the pattern hidden away inside the cerebellum.

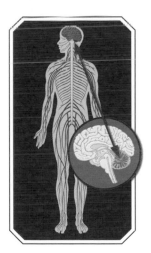

Once thought of as one of the "silent areas" of the brain (which truly exist nowhere), the cerebellum plays a crucial role in controlling physical movement, helping you to balance, to move in a coordinated way, to learn new motor skills like riding a bicycle or playing an instrument. Medieval anatomists knew nothing of these functions but wondered at that little tree drawn into the structure of the cerebellum, and called it the arbor vitae: the "tree of life," the seat of the soul.

As well as this hidden arborescent pattern in the brain, there's another anatomical arbor vitae, inside the cervix of the uterus, where fern-like folds are known as the plicae palmatae ("palm-like folds") or, collectively, as the arbor vitae uteri.

ATLAS

The atlas is the top vertebra of the spine. "Spine" itself comes from the Latin, *spina*. Originally this seems to have meant "thorn" or "prickle," and it then became used for the backbone, too. The English word "spine" retains that double meaning—referring to the vertebral column but also to anything with a needle-like shape. Sticking out at the back of most vertebrae (but not the atlas) is a sharp prong called the spinous process—"spinous" because of its shape, not because it's in the spine. Aulus Cornelius Celsus, a physician who compiled an encyclopedia of medicine in the first century CE, invented the term—or at least, was the first to write it down. Celsus also provided the name for a spinal segment: a *vertebra*. *Vertere* is Latin for "to turn," so perhaps there is a sense here of the whole body turning around the hinge of the spine—the vertebral column as its axis. But individual vertebrae also turn, one upon another, allowing a rotational twist that builds up along the length of the spine. If you're sitting down right now, sit up straight, keep your hips facing forward and twist your shoulders as far left and right as they'll go; allow your head to twist even farther around in each direction. That's some impressive swiveling going on, right there, along the whole length of your spine.

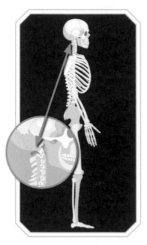

Near the top of the spine is a vertebra with a bony pin that sticks up. It's called the axis, and its pin is the odontoid (Greek for "tooth-like") peg or process—sometimes simply known as the dens ("tooth" in Latin). Unlike other vertebrae, the atlas has no body and is shaped instead like an open ring; the front of the ring articulates with the dens of the axis—and can rotate around it. When you turn your head to the side, a little of this movement comes from a rotation of each cervical vertebra on the one below—but most of it comes from that articulation between atlas and axis: the atlanto-axial joint. So, it's clear why the second cervical vertebra is called the axis.

As for the atlas, it holds up the skull, in the way that the ancient Greek god is often pictured holding up the Earth. But this is all a terrible muddle. In ancient Greek mythology, Atlas held up the sky, not the Earth. The misunderstanding seems to have come from a misreading of ancient sculptures, where Atlas is shown holding up a celestial sphere (perhaps because this is easier than showing him holding up a dome). That sphere is later reinterpreted as the globe. I've compounded the misunderstanding here—with Atlas holding up the Earth.

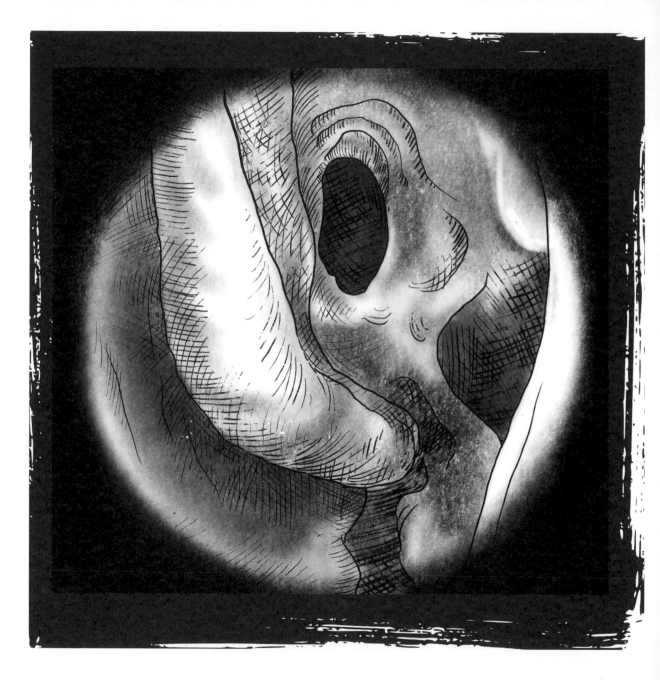

The nose is a much more cavernous place on the inside than it looks on the outside. The nasal cavities, separated in the midline by the nasal septum, are about 3 inches (8 cm) long. They stretch from the nostrils all the way to the back of the hard palate, when they open into the upper part of the pharynx. Each cavity is about half an inch (just over a centimeter) wide at the bottom, narrowing to just a few millimeters wide at the top. And they are about an inch and a half (4 cm) tall, from floor to ceiling. Scroll-like turbinates, or conchae—long curls of bone covered with sticky, mucous epithelium—project into the nasal cavities from the side walls. The two names mean pretty much the same thing. "Turbinate" means something shaped like a *turbo*—a "spiraling shell." *Concha* is the Latinized version of the Greek *konkhe*—also meaning "shell," though originally a bivalve type, like a mussel. (Concha is also another name for the vulva.*) And under these turbinates (or conchae) are narrow openings into more cave-like spaces.

The maxillary antrum is one of these. Or *two* of these, there being a pair, one in each maxilla. The paired maxillae are your upper jaws, bearing all your upper teeth, but they also frame the nasal cavities

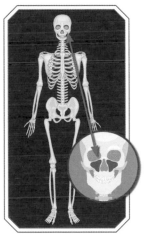

and contain these large spaces adjacent to it. "Maxilla" itself is an interesting word. It's a diminutive form of *mala*, which means "jaw" (and, just to confuse things, "cheek bone"). Perhaps it's because the upper jaw is made of two "little jaws"). Anyway, the maxilla is quite a large bone, and completely hollow, and that hollowness is the pyramidal space known as the maxillary antrum—one of the paranasal air sinuses. Yes, *those* sinuses. The ones that can get infected and become full of mucus and gunk, in sinusitis. *Sinus* is Latin for "hollow curve" and gets used for some large veins as well as these air spaces. Whenever you see "-itis" it signals "inflammation." Actually, the suffix just means "of the"; you're just meant to know that it actually means "(disease) of the ___."

The Latin *antrum* comes from Greek *antron*—a cave. And here it is as seen through the camera of a nasal endoscope.

*Concha also seems very close to the Latin *cunnus* and Old English *cunt*, the author muses (with no sources to back her up). Was female genitalia described as "shell-like?" The other Latin term for female external genitalia, *vulva*, becomes the official anatomical term—and sounds very hard and clinical until you give it a soft "v" and it becomes a "wulwa." Anyway, we're meant to be talking about noses; why are we suddenly off down this genital rabbit hole?

VESICA URINARIA

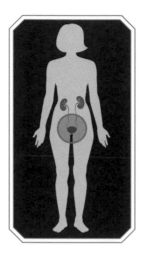

The bladder is one of those organs that very rarely gets described using its proper anatomical name, the *vesica urinaria* (Latin for "urinary vessel"). Even in the technical literature, in English, it's usually just described as "the bladder." That term comes from Old English, *blaedre*, which goes back to an earlier Germanic word, something like *blodram*, which means "inflated." A bladder is something that can be "blown up"—for storage, buoyancy, or perhaps for kicking around. (When Stirling Castle was renovated, a pig-bladder soccer ball was discovered in the rafters of what *might* have been Mary, Queen of Scots' bedroom; she's known to have been very fond of soccer.)

Vesica may seem like quite a boring term, but this beautiful Roman glass vessel reminds us of its origin. And it's related to *venter*, which is Latin for "belly," "bulge," or "womb." That gives us the anatomical term "ventral"—which means "belly-side" (anterior in a human). And it may be that *venter* is a cousin of *uterus* in Latin, both descending from an ancient, Proto-Indo-European (PIE) word for "womb," something like *udero*. Even where *vesica* is dropped from anatomical and clinical texts, it pops up in anything to do with the bladder: the vesical arteries and nerves, vesical calculi (bladder stones), vesical mucus, and the bit where the ureter exits the bladder—the vesicoureteric* junction . . . you get the gist.

Urina is Latin for "urine." Or "urine" is English for *urina*, which is probably the right way around. The Latin comes from the ancient Greek, *oûron*, flowing all the way back to ancient PIE and the word *uer*—meaning to "wet" or "flow."

*I pronounce this "vees-icko-your-ee-terick," but I know some people say "vee-sigh-coh-your-ee-terick." Neither is wrong nor right. But the latter pronunciation, presumably heard by a medical student who'd relied only on their own notes, eschewing the textbooks where such things are helpfully written down, led to a fantastic neologism in an anatomy spot exam. Marking the papers, it took us a while to work out how on Earth the student had arrived at it—they had written: "psycho-ureteric junction." A previously unknown link between mind and ureter.

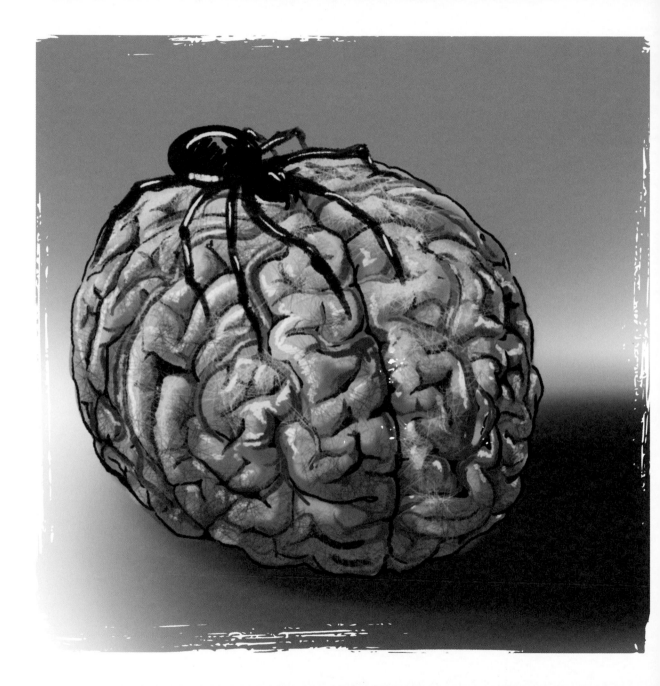

ARACHNOID MATER

L ike something out of a horror film: a brain enveloped in cobwebs.

Your brain is wrapped up in three layers of meninges, inside your skull. The knowledge of these wrappings goes way back: The existence of membranes around the brain is mentioned in the earliest medical text we have, the Edwin Smith Papyrus, from ancient Egypt, and most definitely not written by someone called Edwin. The first mention of the term *meninx* (Greek for "membrane"), which becomes *meninges* in the plural, comes from Erasistratus in the third century BCE, in that bastion of anatomical exploration and teaching, Alexandria. Galen described two layers of meninges, the *pacheia* and the *lepte*—Greek for "thick" and "thin."

The three layers of meninges we recognize today are the pia mater, the arachnoid mater, and the dura mater: the "pious mother," the "cobweb mother," and the "hard mother." Such strange names—poetic but troubling. These Latin names are translations of Arabic words—memorializing the preservation of anatomical knowledge in the Islamic world in the first millennium of the Common Era, after the collapse of the Roman Empire.

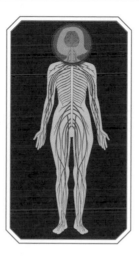

An anonymous Muslim physician described the meninges as *umm al-dimagh*—the "mother of the brain." This seems so odd, but the term described a physical closeness. There is an echo here of the way we talk about "anatomical relations" in English; we don't mean your cousins, we mean things that lie close to each other in the body. The meninges are physically close to the brain. The Persian physician Hali Abbas ('Ali ibn al-'Abbas al-Majusi) followed Galen in writing about two layers, the *umm al-ghalida* ("hard mother") and *umm al-raqiqah* ("thin mother"). When the twelfth-century Italian monk Stephen of Antioch translated these into Latin, they became the *dura mater* and *pia mater* (although *tenue* would have been better than *pia*, retaining that meaning of "thin" and not intruding the slightly odd suggestion of piety). Herophilus had described a cobwebby layer inside the ventricles of the brain, but the seventeenth-century Dutch anatomist (and collector and fabricator of macabre anatomical dioramas) Frederik Ruysch extended the description of the arachnoid mater as the middle layer of the meninges. The term is Latin, from the Greek *arakhnē* and *-oeides*, meaning "cobweb-like."

CAUDA EQUINA

The spinal cord lies in the vertebral canal of the backbone, enclosed by the neural arches of the vertebrae, protecting this precious cable of nerves. Just like the brain, the spinal cord is covered in layers of meninges: The thin, delicate pia mater lies on its surface, while the dura mater forms a looser sheath, with the cobwebby arachnoid mater lining the inner surface of the dura. In between the pia and arachnoid mater is the subarachnoid space—full of cerebrospinal fluid (CSF).

The spinal cord is packed with nerve cells in the center, where they form a grayish butterfly shape in cross section, with nerve fibers stacked up longitudinally around the gray matter, like dried spaghetti, forming the white matter of the cord. The nerve fibers are carrying sensory messages from the body up to the brain and motor messages back down the other way. The spinal nerves leave the spinal cord as separate motor and sensory roots, which then pierce through the dura and unite to form a spinal nerve, which passes through an intervertebral foramen (lined-up notches between two vertebrae) to exit the spinal cord. In the upper

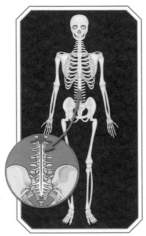

cord, the roots run horizontally. But as you grow, the spinal cord doesn't keep up with the dural sheath or the spinal column itself. Your backbone finishes down at your tailbone or coccyx, but—as an adult—your spinal nerve finishes level with the first lumbar vertebra, just below the bottom of your ribcage. This means all the lumbar and sacral nerve roots have some way to travel down inside the vertebral canal—and keep traveling downward (or caudally) once they've escaped their bony confines. The dural sheath extends down into the sacrum, and it's expanded at the bottom end, below the cord, forming the lumbar cistern (from Latin *cisterna*, for a subterranean water reservoir), full of CSF—and all those lumbar and sacral nerve roots. In the dissection room, you can carefully remove the neural arches of the vertebrae, at the back, and then you can see the dural sac lying in the vertebral canal. Taking a scalpel and precisely incising the dura, lengthways, you can open it out and, inside, find the spinal cord—and below its tapered end, a thick skein of nerve roots. They look a bit like a horse's tail, and that's their name, in Latin—the *cauda equina*.

GLOMERULUS

Now we're in the kidneys and looking at something which is just about visible to the naked eye: a tiny bobble that's about a third of a millimeter across. Under the electron microscope, the true nature of the tiny tuft reveals itself—and it's a knot of minute capillaries. This is where the business of the kidney happens—the capillary knot is fed by blood from a small arteriole, and the sheer pressure of the blood pushes fluid out of the capillaries into the surrounding space, which is known as Bowman's capsule. This fluid is "raw urine"—it will pass along a looping system of tubes where some substances are reabsorbed into the blood and more waste products are pulled out into the urine. It's all very finely balanced to make sure that the blood that leaves the kidney is of a perfect composition. The urine that the kidney makes collects in its interior, then makes its way down to the bladder via the ureters, ready to be gotten rid of at a convenient moment.

The knot of capillaries where the work of the kidneys begins is called a *glomerulus*, the diminutive form of the Latin *glomus*, a ball of yarn—but it's only been known by this name since the nineteenth century.

The glomerulus with Bowman's capsule around it forms a structure known as a renal corpuscle. In the fourth century of the Common Era, the Byzantine

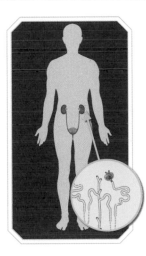

physician Oribasius described, in Greek, *ta sōmata tōn nephrōn*—which becomes *corpusculum renalis* in Latin, renal corpuscle in English. By the seventeenth century, anatomists had microscopes to extend their observational abilities, and the Italian anatomist Marcello Malpighi described the structure of glomeruli inside the renal corpuscles, though he did not yet use this name for them. He wrote that these little blood-vessel balls looked like "fruit hanging from a branch." Following on from his work, these minute but important structures were variously referred to as Malpighian corpuscles, bodies, coils, or tufts. In 1782, Alexander Schumlanski wrote his doctoral thesis on the structure of the kidneys, in which he mentioned *glomeres* as a term. In 1841, a young English surgeon by the name of William Bowman described the detailed structure of Malpighian bodies and ended up bequeathing his name to the capsule around the glomerulus. Together with the physiologist Robert Bentley Todd, with whom he trained, Bowman is credited with discovering the connection between the capillary knots and the urinary tubules—though in fact, Schumlanski had noted that connection more than three quarters of a century earlier. And it was a German surgeon and anatomist, Wilhelm Busch, writing about the excretory system of snakes in 1855, who finally gave the world *glomerulus*.

CORACOID PROCESS

Another "oid" word. Once again the "oid" comes from Greek, *-oeidēs*, meaning "form." And here the "corac-" part comes from the Greek *korax*, meaning "raven"—surely an onomatopoeic name for this bird. (In Latin, *corax* also spills out to mean "battering ram"—taking its cue from the powerful beak of the raven).

This beak-like prong of bone projects from the upper, outer edge of the scapula, pointing forward. Wherever you see a projection of bone like this, it's odds-on that it's there for muscle attachment, forming a lever for muscles to pull on. The coracoid process, though quite small, has three muscles attaching to it. One of them takes its name from the process: the coracobrachialis muscle, stretching down to attach to the humerus in the arm (*brachium* being "arm" in Latin, *brakhion* being "arm" in Greek). The other two are the short head of biceps* brachii muscle (literally: the two-headed [muscle] of the arm) and pectoralis minor (literally, the small [muscle] of the chest).

The coracoid process also helps to stabilize the lateral end of the clavicle, which sits just above it, reaching across to form a joint with the uppermost, outermost

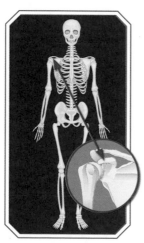

projection of the scapula: the acromion (which you may remember from page 14). A strong coracoclavicular ligament binds the lateral clavicle down to the coracoid process beneath it, stabilizing that acromioclavicular joint.

You can feel your own coracoid process. Push into the slight cavity just under the clavicle and medial to the shoulder. Then walk your fingers farther laterally, to the side, and you should be able to feel a distinct knobble of bone. That's it. Your very own little pet ravens, one on each side.

*Please note: "biceps" is singular; don't be fooled by the "s." There is no such thing as "bicep."

MENISCUS

The knee is a deceptively simple-looking joint. Two bones hinging against each other, with a small bone helping to provide the main extensor muscle with a bit of mechanical advantage—moving its angle of pull away from the joint. But in fact, it's not just a hinge, because there's a bit of axial rotation going on in there, too, and some forward–backward sliding between the two bones as well. In a simple hinge joint, the two joint surfaces can fit snugly together. We see this in the joint between the humerus and the ulna, for example. But if you need to accommodate other movements, the fit has to be looser. This creates a problem for lubricated, synovial joints, when you only ever want a thin film of fluid between the joint surfaces—but the solution is to introduce a washer into the joint. In some joints, the washer might completely separate the two articular surfaces—as in the joint at the inner end of your clavicle, the sternoclavicular joint. But in others, it's an incomplete disc—like the menisci in the knee.

In the nineteenth century, the menisci were thought to be vestigial structures. But now we know they're essential to knee function. They're made of fibrocartilage, and they're crescent-shaped looking

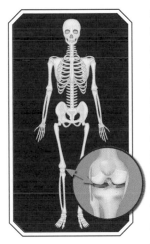

down on them from above, as in the facing image, but they have a triangular cross section—thick at the outer edge, very thin on their inner curve. They're important in transmitting load across the knee joint—they help to spread the load across the femoral condyle. If a meniscus is removed, only a small part of the femoral condyle is left in contact with the tibial articular surface—with resulting high stress on that area. The menisci also help with shock absorption and maintain the congruity—or close fit—between the moving parts of the knee, allowing sliding and rotating movement to happen, but not *too much*.

Meniscal tears are quite common injuries, happening when the knee twists too hard, and are often accompanied with ligament injuries, such as torn cruciate ligaments. In the 1970s, the standard approach to badly torn menisci was to whisk them out, as they were still viewed as vestigial. Now that we know how useful menisci are, surgeons will try to patch them up if they can, and there's ongoing research into replacement implants.

The word "meniscus" is Latin, coming from the Greek *meniskos*, the diminutive of mene—"moon"—and so it means "mini moon" or "crescent."

PIRIFORMIS

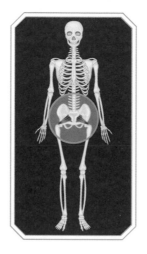

You may have heard the idiom that "it's all going a bit pear-shaped." It means that something's gone awry, and it's said to derive from a Royal Air Force expression for less-than-perfect aerobatic circles, but this may be apocryphal. During the Second World War, various bombs and mines were also described in the papers as "pear-shaped." The term can also be used to describe someone's figure, relating to an unfortunate accumulation of fat around the hips. But the precise origin of the idiom remains obscure.

But the pear-shaped muscle in the buttock has a much longer history, with the name *pyriformis* (Latin for "pear-shaped") going back at least to the seventeenth century. (There are a couple of other, smaller, pear-shaped pieces of human anatomy: the piriform recesses lying alongside the epiglottis in the throat, and a small piriform lobe on the undersurface of the brain.)

With a wide (but flat) belly attaching from the middle three segments of the sacrum, the piriformis muscle passes out of the pelvis through the greater sciatic foramen, thinning into a slender tendon that inserts into the greater trochanter of the femur. When piriformis contracts, it rotates the thigh outward.

The sciatic nerve emerges from the pelvis into the buttock, between piriformis and the sacrospinous ligament just below. There's a condition called piriformis syndrome, which is linked to sciatica, with a suggestion that the muscle somehow compresses or irritates the nerve. But it's a controversial diagnosis, as the symptoms are so similar to those experienced with compression of nerve roots in the lumbar spine, due to intervertebral disc prolapse. But it does seem to be responsible for potentially one in twenty cases of sciatica, and can be, quite literally, a pain in the butt.

"Muscle" is an intriguing term, too. It comes from *musculus*, the Latin for "little mouse." Perhaps the movement of muscles under the skin looks a bit like a mouse darting around under a rug. Or perhaps, when a small muscle with a long tendon—like any number of the forearm muscles—is dissected out, it could be thought to look a bit like a little pink mouse with a long white tail.

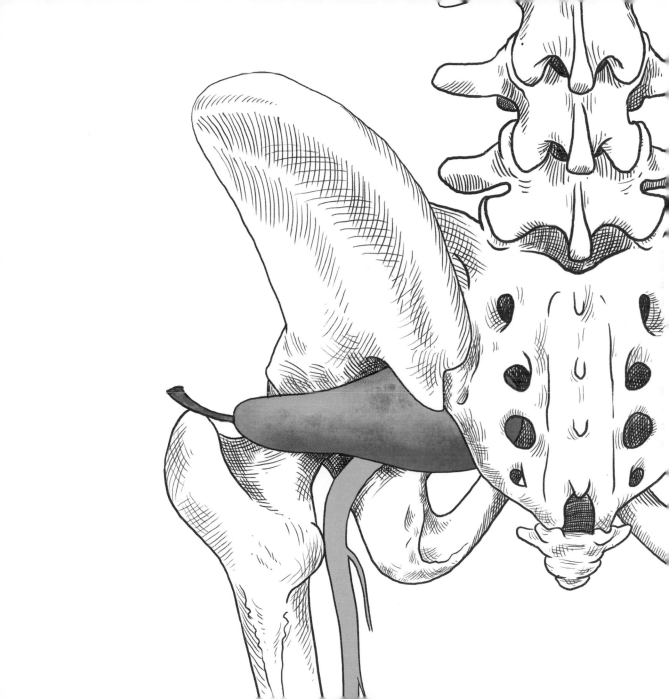

VESTIBULE

"Vestibule" comes from the Latin, *vestibulum*, meaning an entrance hall or lobby. And in fact it carries the meaning of "cloakroom"—a place to put on and take off your clothes, or *vestes*. It's used anatomically to describe all sorts of openings into larger or longer spaces. There's the vestibule of the vagina, which is the space between the labia minora in the vulva, containing the openings of the vagina and urethra. There are vestibules in the atria (*atrium* being another name for a room in Latin) of the heart, just above the mitral and tricuspid valves. There's a vestibule in the inner ear, a bony cavity lying between the semicircular canals and the cochlea. There's a vestibule in the larynx, just under the epiglottis and above the vocal cords. There's a vestibule in the mouth: the space between your lips, cheeks, and teeth—it's the space your toothbrush is in when you're brushing the outer surfaces of your teeth. And there's a nasal vestibule, shown in yellow on this psychedelic image of the insides of your

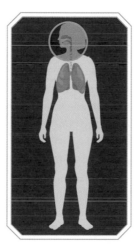

nose (which also shows nicely how long and high the nasal cavity is).

The nasal vestibule is just inside the nostrils, and its border is marked by a curved ridge called the limen nasi—*limen* being Latin for "threshold" or "doorway." The vestibule is lined with the same sort of skin as on the outside of the nose, and just inside the nostrils, this skin bears hairs to trap unwanted particles. Above and behind the limen nasi, the lining changes to a ciliated epithelium (see page 77) with plenty of mucus-producing goblet cells. This is all about cleaning the air you breathe in—creating a sticky film of mucus inside the nasal cavity to trap particles. The tiny, hair-like cilia beat the mucus to the back of your nose—moving it 6 millimeters a minute—where it can drop down into your pharynx to be swallowed. How disgusting! You're eating snot all the time. If you're producing a lot of mucus, some will end up coming out the other way, too, through the vestibules and dripping out of your nostrils.

COCHLEA

Curled up tight inside the hardest part of your skull is the cochlea—from the Greek *kokhlias*, "snail." Just under a centimeter in diameter, this is the engine of hearing, transforming vibrations in fluid into electrical activity—and that signal is conveyed to the brain via the cochlear nerve.

In the anatomy department where I used to work, we had boxes of temporal bones, the bones that house the workings of the ear. But the bones were like boxes, too. Someone had carefully sawn them open and then put them back together with tiny hinges and a hook to close them. When you opened them up, the treasures inside were revealed: the semicircular canals and a cross section through the minute spiral of the cochlea. With all the twists and turns in these cavities, it's not surprising that the collective term for them is the bony "labyrinth." This name was given to them in the seventeenth century, with anatomists

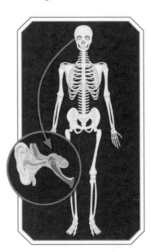

following the tradition of naming parts in Latinized Greek. The Greek word meant "maze," too, but may have an even deeper meaning. It seems to be related to *labrys*, a word for a double-headed axe. Such axes are symbolically linked with Zeus—and, on Crete, with goddesses and the royal palace of Knossos. If the palace is the "Place of the Double Axe"—then its elaborate architecture could have given *labyrinthos* a new meaning. But there are no Minotaurs in the bony labyrinth, just a tiny snail.

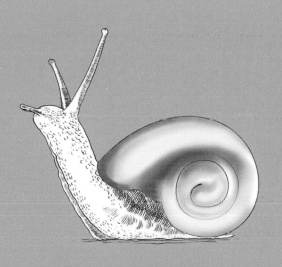

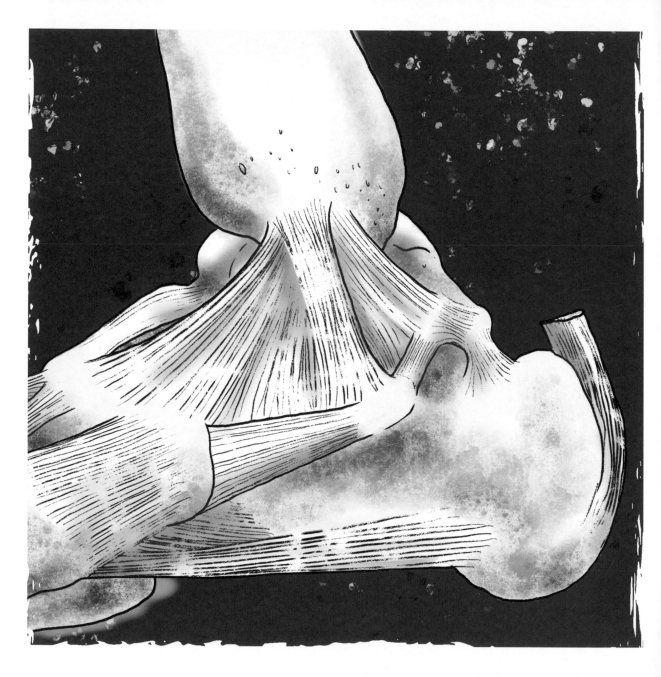

LIGAMENT

The Latin verb *ligare* gives us the verb "to ligate," which is used to describe the act of surgically tying up something, usually an artery, to stop it bleeding. It also transforms into the noun *ligamentum*, meaning "a binding" or "a bandage." And this is adopted as the anatomical and quite familiar term for ropes, sheets, and bands of tough fibrous tissue that bind bones together, supporting joints. Ligaments are similar to tendons and fascia—they're all fibrous connective tissue. Fascia is everywhere in the body—it wraps up and binds everything from organs and muscles to blood vessels and nerves. Almost synonymous with the original meaning of *ligamentum*, the Latin word *fascia* also meant "a bandage" or, perhaps, "a belt." Just to confuse the issue, some condensations of fascia in the abdomen and pelvis are also called ligaments—and folds of peritoneum are, too—and so we have the uterosacral ligament, the lienorenal ligament, the triangular ligaments of the liver, the phrenicocolic ligament. None of them have anything to do with bones. The ligaments

that attach bones together are stronger and denser, and they need to be. They're also slightly stretchy, which allows them to absorb strain. They are made up of long fibrils (from the Latin, *fibrilla*, "little fiber") of the protein collagen, with the cells that manufacture this protein, fibroblasts, squished in between them. (Collagen comes from the Greek, *kolla*—"glue"—and -*gen*, "making"; animal glue is made from boiling up skin, bone, tendons, and ligaments).

The ligament shown here, on the inside of the ankle, is the deltoid ligament, also known as the medial collateral ligament. It is triangular—shaped like a Greek capital delta, Δ. It's made up of several discrete bands, attaching from the malleolus (knobbly bit, literally "little hammer") of the tibia down to the talus and calcaneum. The deltoid ligament can be damaged in sprains where the foot turns outward—eversion sprains—but it's not as commonly injured as the ligaments on the outer side of the ankle joint, which form the lateral collateral ligament, often torn in inversion injuries.

MAMMILLARY BODIES

Landscapes are full of lumps, bumps, hills, and mountains named after a particular part of female anatomy, from the Paps of Jura to Mamelles Island in the Seychelles, from Mam Tor in Derbyshire to Grand Teton in Patagonia. The landscape of anatomy is also littered with mammary allusions.

I have an impression of early anatomists, almost exclusively men, poring over the intricate structures of the human body and becoming quite excited when they found anything that reminded them of a bit of female anatomy. It's extraordinary how many parts of the body, apart from the breasts themselves, are named after breasts and nipples.

The adjective "mastoid" comes from the Greek *mastoeidēs*, meaning "breast-like." Just below your ear, there's a prong of bone sticking down, forming an attachment point for a long, strap-like muscle in the neck. This pyramidal lump of bone is called the mastoid process. (And the long muscle is called sternocleidomastoid, because it attaches from the sternum and clavicle—*cleidos*, in Greek—up to the mastoid process.)

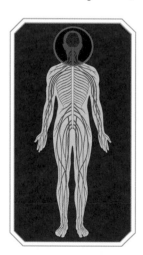

The adjective "papillary" comes from Latin *papilla*, meaning "nipple" or "teat." There are papillary muscles in the heart, holding down the chordae tendineae (tendinous cords—or heartstrings), which stop the valves turning inside out. The duodenum contains a papilla where the bile duct and pancreatic duct open. And the tongue is covered with tiny papillae, bearing taste buds.

The adjective "mammillary" comes from Latin *mamilla*, meaning "little breast" or "nipple." There's a pair of mammillary processes sticking out of the back of the articular processes of each lumbar vertebra. And the two rounded mammillary bodies highlighted in the picture here form part of the hypothalamus of the brain.

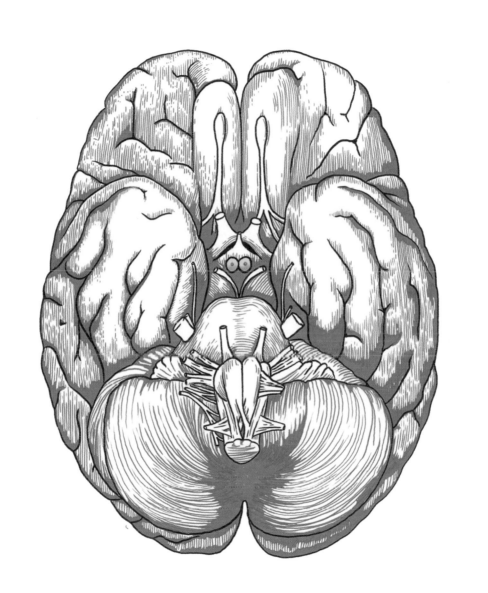

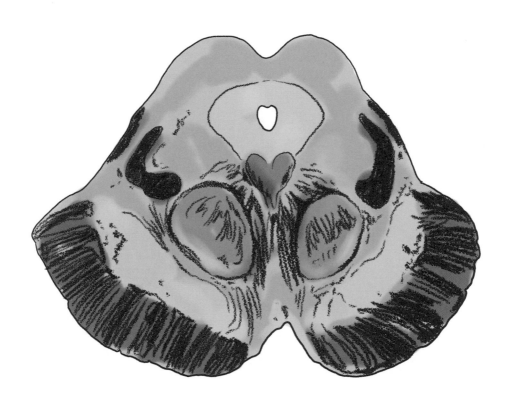

TECTUM

The cerebrospinal fluid that is made in the ventricles inside the brain flows down through a narrow channel, the cerebral aqueduct, into the fourth ventricle, where it escapes out through the foramina of Magendie and Luschka. The cerebral aqueduct passes through a part of the brain called the midbrain. This term is not too much of a puzzle. The midbrain sits between the forebrain and the hindbrain. This all harks back to the development of the brain in the embryo, when three bubble-like vesicles (from the Latin for "little bladder," *vesicula*) form at the front end of the neural tube (the precursor of the brain and spinal cord).

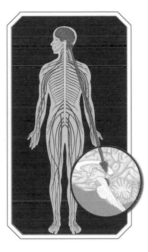

But this is anatomy, so there's always a handy Greek word if you don't like the English terms. The midbrain becomes the mesencephalon, sitting between the prosencephalon and the rhombencephalon. *Enkephalos* is Greek for "brain" (literally—the "in-head"), *proso-* means "forward" and *meso-* means "middle." *Rhomb-* is odd, though. In Late Latin, it came to mean the geometric shape we might understand by it today, a rhombus—a quadrilateral with equal-length sides. But in earlier times, this Latin word meant "magician's circle" (a shape used in magical rituals) or . . . a fish—specifically, a turbot. The history of words can be so weird.

The Latin word derived from Greek, *rhombos*, and this meant a spinning thing—from *rhembo*, "I turn." And specifically, it was used to refer to a type of whirling instrument—a bullroarer. These are made of a slender slat of wood or bone, tied to a cord, and whirled around to create a roaring noise. Some bullroarers are made with a pointed piece of wood—is it this that provides the evolution of the shape implied by the term "rhombus," from circle to diamond? Anyway, this is largely academic as "rhombencephalon" is a late addition to the anatomical lexicon, introduced by Wilhelm His in 1890, when he used it to describe the cerebellum together with the lower medulla oblongata. If you cut the cerebellum away from the brainstem, you're left with a *diamond*-shaped opening into the fourth ventricle.

Putting these musings on the rhombencephalon to one side, and focusing back on the midbrain, where we are meant to be directing our attention: The cerebral aqueduct (a great name that needs little interpretation) is roofed by the tectum and floored by the tegmentum. Although *tectum* in Latin can mean "roof," it comes from the verb *tego*, "to cover." And guess what—so does *tegmentum*. Nowhere are the twists and turns, the spinning and whirling of nomenclature and etymology, so tortuous as in neuroanatomy.

OLECRANON

This is the posh name—the correct name—for the knobbly bit of your elbow. It comes from Greek, *ōlene*, meaning "elbow" or forearm, and *kranos*, meaning "helmet." (The very similar *kranion*, meaning "skull," passes into Latin as *cranium*.) The "helmet of the elbow" makes a lot of sense—that large knob of bone offers some protection. But actually it's like any other prong or process of bone: It's principally there as a lever for muscle attachment. This is where triceps brachii attaches, and the muscle pulls on this lever to extend or straighten the elbow. Triceps brachii is the three-headed muscle of the arm, sitting around the back, while biceps brachii, the two-headed muscle, sits in front. Two of the heads of triceps, the medial and lateral, arise from the humerus, while the long head originates from the scapula, just below the shoulder joint, on the almost completely inconspicuous infraglenoid tubercle.

The socket of the shoulder joint, on the scapula, is called the glenoid fossa, and this is a curious term indeed. *Fossa* is straightforward—Latin for "ditch," from the verb *fodere*, "to dig" (which also gives us a term for something that is dug up: "a fossil," and for animals that dig tunnels: "fossorial"). But the Greek word *glēnos*, popping up in Homer, means "a mirror."

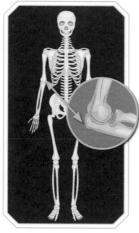

A similar word, *glēnē*, means "eyeball" (does the eye mirror the world for us?), and then, perhaps by extension, the eye socket as well. The Roman physician and anatomist Galen was the first to use "glenoid" to describe the joint of the scapula, but nobody knows whether he was going for the meaning of "mirror" or "socket." Perhaps it was both, with the shiny articular cartilage of the glenoid fossa giving the socket a glassy appearance.

The olecranon is part of the ulna, one of the two forearm bones. *Ulna* is Latin for, well, "the ulna," and also comes from the Greek, *ōlene*. So if you decode "the olecranon of the ulna" you get the helmet of the elbow of the elbow. Which is confusing to say the least.

The accompanying *radius* in the forearm is a Latin word, meaning "staff," "spoke," or "weaving shuttle." An old Greek name for the bone, *cercis*, also means "weaving shuttle." I'm left wondering whether this relates not only to the long, slender shape of the bone, but to its ability to rotate—to "shuttle around"—the ulna, as you swivel your hands to face one way then the other. But maybe there's another explanation—animal limb bones were used for weaving in antiquity, so perhaps the bone was called a *cercis* or *radius* not just because it looked like a weaving shuttle—but because it *was one*.

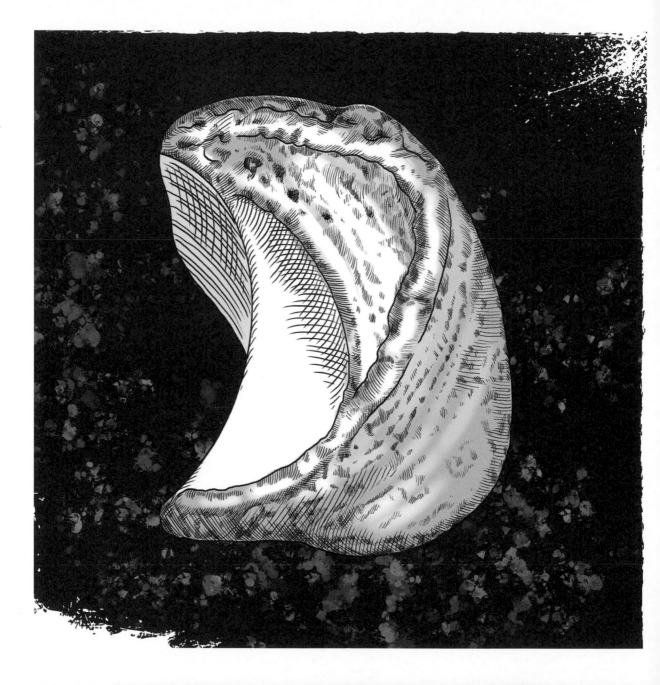

LUNATE

nother crescent-shaped piece of anatomy, this time from the wrist. The lunate is one of the carpal bones, *carpus* being "wrist" in Latin, from the Greek, *karpos*. The word is suggested to have come from an ancient Proto-Indo-European root, **kwerp*, meaning "to turn" or "revolve"; the wrist is a very mobile joint and allows you to swivel your hand around. Somehow that same old word passes down other branches of the Indo-European family tree to become *wreathe*, *writhe*, and *wrist*.

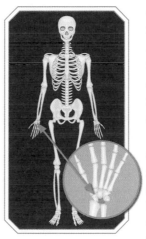

There are eight bones in the human carpus, in two rows: the scaphoid, lunate, triquetral, and pisiform in the proximal row and the trapezium, trapezoid, capitate, and hamate forming a distal row. They're all named for their shapes. We've already met the hooked hamate and pisiform (forming the medial boundary of Guyon's canal). Triquetral comes from Latin, *triquetrus*—"triangular" or "pyramidal." The trapezium and trapezoid are both sort of quadrilateral, from Greek *trapezoeides*, "table-shaped." The capitate comes from Latin, *capitatus*, "with a head"—and it does look very much like a little knob-headed chess pawn, sitting upside down in your hand as you look at it. The scaphoid comes from Greek, *scaphoeides*, meaning "boat," and it does look a little bit like a tiny coracle. And when you turn the lunate around and look at it from the side—it is a tiny crescent.

Of all the carpal bones, only the scaphoid and the lunate are involved in the wrist joint proper, nestling in and articulating with the shallow bowl-shape at the end of the radius. The others are all, quite literally, hangers on—bound to the scaphoid and lunate via numerous ligaments. When you bend your "wrist," the movement happens not just at the anatomical wrist joint, the radiocarpal joint, but between the two rows of carpals, too, adding extra range to flexion and extension.

The very mobile human wrist joint is a legacy from your arboreal primate ancestors, who liked to hang around in trees, and for whom being able to grab branches in a variety of orientations was a useful survival skill. You might not hang around in trees much anymore (though my children do) but a mobile wrist joint has proven itself to be an incredibly useful thing to have for a species that habitually makes and uses tools.

LOOP OF HENLE

The nineteenth century saw significant advances in light microscopy—in particular, the development of lenses that reduced the scattering of light of different colors—opening anatomists' eyes to an even greater level of detail. A German physician and anatomist, Jakob Henle, made the most of the new technology, driving forward the new science of histology. This is the study of body tissues, from the Greek *logos*, literally "word," but with a much wider meaning, "study," and *histos*, meaning "web" or "woven." Presumably the German anatomists thought this was the best Greek word they could find to describe something that the ancient Greeks didn't actually have a word for.

Henle delved into the histological detail of the body, uncovering new insights into the formation of mucus and pus, the lymphatic system—and the kidney. He also brought a systematizing approach into histology. Where other scientists were racking up more and more disparate types of tissues, Henle recognized four basic tissue types, combined in different ways in the body's organs: epithelial, connective, muscular, and nervous tissue. Henle did so much pioneering histological work that his name became associated with structures all over the body, from the crypts of Henle in the eye to Henle's fissures in heart muscle, from Henle's ampulla

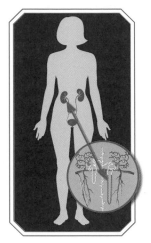

in the oviduct to Henle's ligament in the abdominal wall. But perhaps the best-known of these eponyms, and the one which has most fiercely resisted replacement with a more scientific term, is in the kidney, where the nephron loop is still often referred to as the loop of Henle. This loop is part of the renal tubule, which starts at Bowman's capsule, where fluid is first filtered from the blood. The renal tubule itself is 1 to 2 inches (3–5 cm) long (and there are around a million of them in each kidney) and comprises a proximal convoluted tubule, a loop of Henle, and a distal convoluted tubule. The fluid entering the loop becomes more concentrated as it runs down it, as water is reabsorbed into the blood, and then more diluted again as it ascends back up the loop. By the time the fluid leaves the loop, 20 percent of its original water and 25 percent of useful ions like sodium and chloride have been reabsorbed back into the blood. The kidney is constantly working not just to remove waste products from the blood, but to keep the water and sodium balance of the blood *just right*. You don't have to think about this—and are only really aware that your body is making urine when enough of it collects in the bladder for you to consider emptying it. But the kidneys, and those two million loops of Henle, are busy all the while. Even while you're reading this.

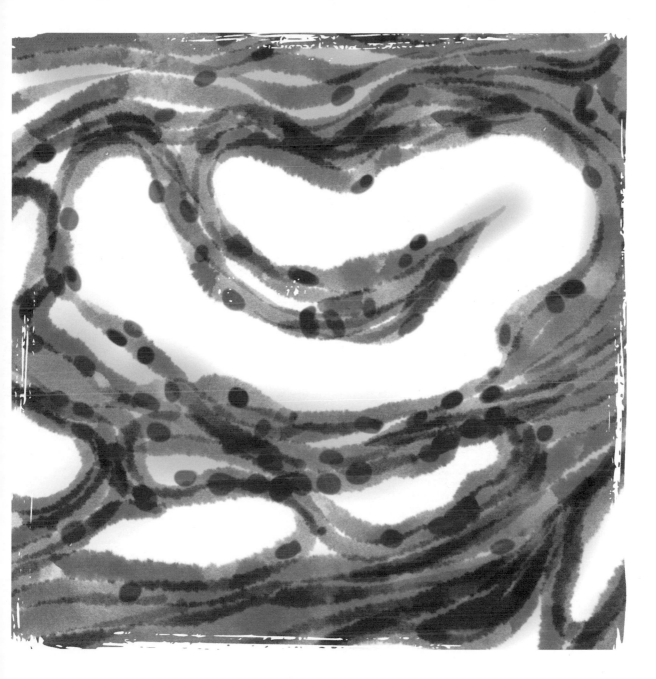

SPROUTING AXON

This is a nerve cell attempting to repair itself. The technical word for a nerve cell is a neuron, which is a bit annoying as *neuron* actually means "tendon" in Greek. Oh well, they can look a bit similar in the dissection room, where some tendons appear long, slim, shiny, and white, with the impression that they are made of smaller fibers stacked up. Nerves are not *quite* as shiny, though, and their make-up is completely different. Whereas tendons are made of aligned collagen fibrils, nerves are made of assembled nerve fibers—the long extensions of neurons—also called axons. That's an odd word, arriving into anatomical nomenclature in the nineteenth century; it's Greek for "axle." And the word goes right back to Proto-Indo-European, something like *aks*, which also gives us "axis." (Proto-Indo-European was a Bronze Age language, in existence around 4500 to 2500 BCE, containing many words for domesticated animals as well as wild ones—and this word *aks*, which relates to the axle of wagons; the wheel was invented hot on the heels of horse domestication in the early Bronze Age.)

Some axons are extremely long. The cell bodies of upper motor neurons reside in the primary motor cortex of the brain. (*Cortex* is Latin for "bark" and is

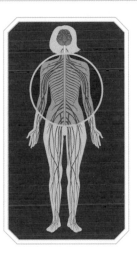

used to describe the outer layers of various organs, including the brain.) Some upper motor-neuron axons travel all the way from the cortex down to the bottom of the spinal cord, before they synapse (*synapsis* is Greek for "connection") with lower motor neurons. Doing a rough measure on myself, from the top of my skull down to my first lumbar vertebra, that's about 28 inches (70 cm)—which is pretty long for just one cell. But the lower motor neurons send their axons traveling out in spinal nerves, which break up into peripheral nerves, some traveling all the way down to innervate muscles in the feet. That means a single axon could reach all the way from the end of your spinal cord almost to your toes, and that's about 50 inches (130 cm) on me. Impressive.

But if those nerves are severed, or a disease leads to dieback of axons, the connection between your brain and your muscles is lost and weakness or complete paralysis (if enough axons are lost) ensues. The cut ends of the axons will attempt to regenerate, sprouting just like a coppiced hazel. Drug treatments can be used to stimulate this regrowth, to support recovery in peripheral nerves, and hopefully, eventually, within the brain and spinal cord as well.

You may look at this picture and think, "That's just an eye." And that's true, but anatomical terminology is all about precision. If we just use the word "eye"—what are we talking about? The part of the eye that is visible from the outside, or the whole eyeball (technically, the orbit)? And in fact, what the term "palpebral aperture" denotes is not really the eye at all, but just the opening through which we see the eye, and through which the eye looks out on the world.

"Aperture" comes from the Latin, *apertura*, "an opening," from the verb *aperire*, "to open" or "uncover." (This is thought to come from the Proto-Indo-European *ap-wer-yo*, meaning "away cover"—which makes me think of a magician whisking away a tablecloth from under a tea set.)

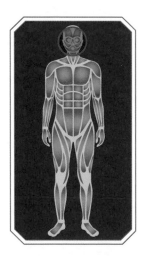

Palpebra is "eyelid" in Latin—which is related to the verb *palpare*, to "touch" or "stroke," which also gives us the word "palpate," which is used to describe the hands-on medical examination of a patient's body. Perhaps the eyelid gently strokes across the surface of the eye. But it also reminds me of a butterfly kiss—a gentle kiss with the eyelashes.

CILIATED EPITHELIUM

One of Jakob Henle's fundamental tissues of the body, epithelium comes in a wide variety of types, all of it lining surfaces—either of the outside of the body or of the various tubes and cavities inside it. And here the attack of the nipples continues, as "epithelium" comes from the Greek, *epi*, meaning "upon" and *thēlē*, "teat." Well, yes, nipples are covered in epithelium, but so is the rest of your body. And the inside of your body. This is really starting to feel like an obsession now.

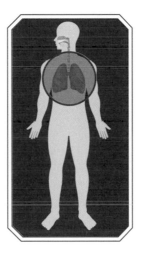

Epithelium can come in simple sheets, just one cell thick, sitting on a basement membrane. This is quite a strange term in itself, but the sense is that the membrane is "base-forming" rather than something to do with cellars. Stratified epithelia consist of layers of cells on a basement membrane; "stratified" coming from Latin, *stratum*—"a spread-out thing," from the verb *sterno*, "to spread" or "stretch out." A *stratum* could be a word for a generically spread-out thing, a blanket, or—even more specifically—a saddle cloth. Epithelia are also categorized by the shape of cells in them: cuboid (a familiar word from the Greek for "die," *kubos*), columnar (tall and thin, from Latin *columna*, "column" or "pillar"), or squamous (a less familiar word, but a very useful one—from the Latin *squama*, meaning "scale").

Finally, epithelia can be categorized according to what they make, what their surface is like, and where they sit in the body. Respiratory epithelium lines the respiratory tract, from the nasal cavity—behind the limen nasi—all the way down into the bronchi and bronchioles inside the lungs. Respiratory epithelium is mostly columnar (and pseudostratified, as it can *look* like it's made of layers down the microscope, even though it's really only one cell thick) but becomes very thin—comprising squamous cells—in the bubble-like alveoli of the lungs. But two very important things about respiratory epithelium are that it contains goblet cells, which produce mucus, and that it is *ciliated*. On its surface, it bears a fringe of minute, hair-like cilia (a Latin word that actually means "eyelash"), which move to waft the mucus up and out of the lungs, once again taking it to the throat to be swallowed. Delicious.

URETHRA

The urethra is the tube through which urine flows out of the body. It's about an inch and a half (4 cm) long in a female and about 8 inches (20 cm) long in a male (though this is somewhat variable). The female urethra is embedded in the anterior vaginal wall—and if the fascial supports of the vagina are damaged in childbirth, this can contribute to incontinence. The bottom end of the urethra opens into the vaginal vestibule about a couple of centimeters behind the clitoris. The male urethra passes through the

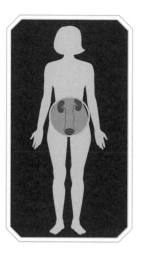

prostate gland, and so enlargement of the prostate can cause problems with urination. Inside the prostate, the male urethra is joined by the ejaculatory ducts, formed by the union of the seminal ducts and the vas (or ductus) deferens—so the urethra in the male is the way that semen, as well as urine, gets out of the body.

Urethra is a Late Latin word meaning . . . "urethra." It comes from the Greek, ourethra, mentioned by Hippocrates as a passage for "urinating"—the Greek verb ourein being similar to the word for "to water" or "to rain": huo. Confusingly, the word "ureter" originally meant the urethra, too; Hippocrates and Aristotle use the two terms interchangeably. The Greek word oureter is also formed from the verb ourein—and literally means "the urinator."

It was around the year 40 CE that the differentiation was made—by the Greek physician Aretaios, who was keen on precise definitions: The ureter was the tube carrying urine from the kidneys to the bladder, while the urethra conveyed urine from the bladder to the outside world. But "urethra" as a designated term for the lower tube was slow to catch on. Galen wrote about the "urinary channel" and "neck of the bladder," but didn't name the urethra. The principal translator of the ancient medical texts into Arabic, in the ninth century, Hunayn ibn Ishaq, followed Galen in using a term that just meant "urinary channel." In the eleventh century, when European medical schools were getting going again, the Arabic texts were translated into Latin and the urethra appears as the porus uritidis, the "urinary channel," or collum/cervix vesicae, the "bladderneck." And then Jacopo Berengario, writing his own anatomy textbook in 1521, further muddied the waters. "The canal of the penis," he wrote, "is called by some the uretra." Aretaios must have been turning in his grave. Eventually the mess was sorted out, with the simple Latinization of the Greek word, into urethra, and the centuries of confusion were forgotten. Water under the bridge.

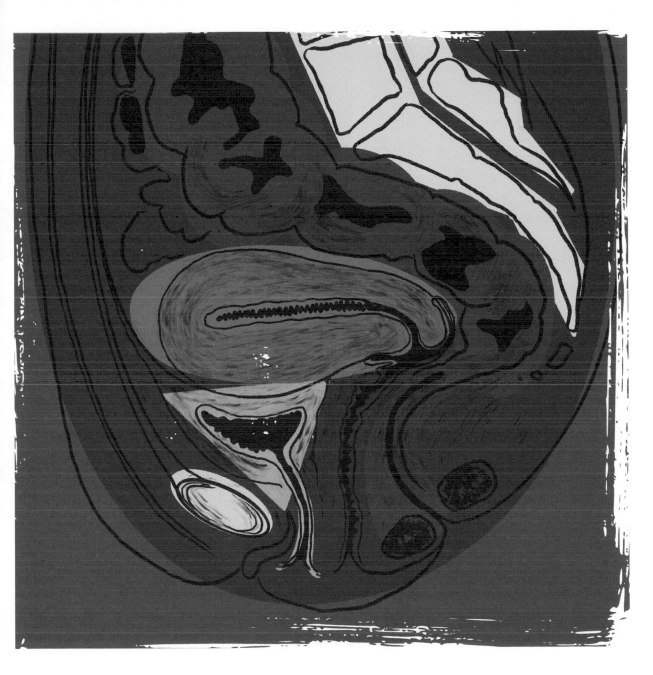

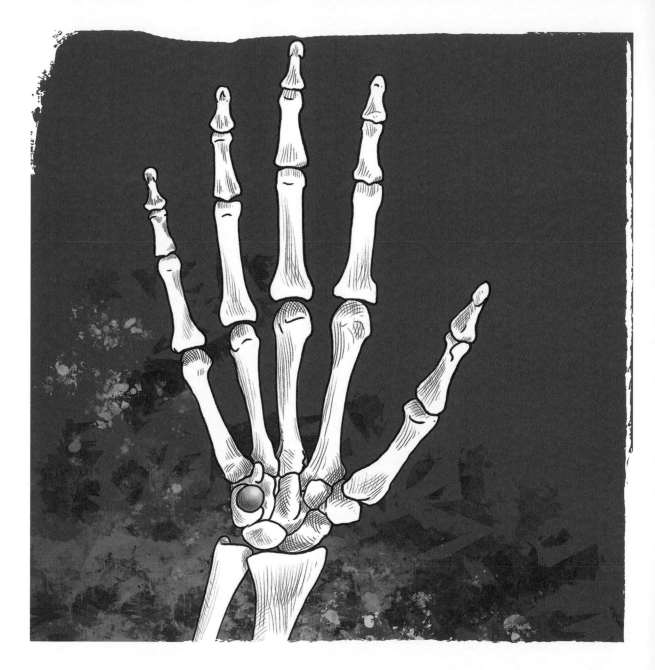

PISIFORM BONE

One of the eight bones of the carpus, the pisiform really has no business being a carpal bone. It doesn't sit neatly in the first row of these bones; it's stuck on the front of the triquetral bone. All the wrist bones start off life preformed in cartilage. By six years of age, the rest of these cartilage models are well on the way to becoming bone—but not the pisiform. It doesn't start ossifying until its owner is eight to twelve years old. It's also closely related to—ensheathed by—the tendon of one of the forearm muscles, flexor carpi ulnaris. All these things mark out the pisiform as *different*.

Is it really just a tendon-bone—a sesamoid bone that develops in a tendon? Is it an anatomical memory, a vestige of a hand that once had a sixth finger? Or is it a vestige of a once more-complicated set of carpal bones, a remnant of a central row that has otherwise disappeared? This last suggestion is perhaps most likely.

Such questions exert the investigative faculties of comparative anatomists and evolutionary morphologists. We like to know not just what things are now, and what they do, but where they *came from*.

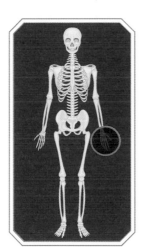

What everyone agrees on, though, is what its name means—and that is "pea-shaped," in Latin.

Beyond the carpals are the metacarpals, and then there are the phalanges—two in the thumb, three in each finger. The singular, "phalanx," is a Greek word (no surprises there) and seems to have been coined—or at least, first written down—by Aristotle. It's always said that he was thinking of a little row of soldiers when he chose this word, and *phalanx* is indeed the word for a line of battle. But it also means "spider"—and so I wonder if he really had that in mind—as my fingers scuttle over the keyboard, tracking down their etymological prey.

ORBICULARIS OCULI

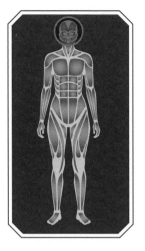

Like some kind of incantation, orbicularis oculi weaves its spell, and it's a sleeping spell: Just relax and close your eyes. The outer fibers of this muscle lie beyond the eyelids, spilling out onto the forehead and cheekbone. Each fiber starts from an attachment to the bone on the medial, inner side, orbits around and circles back to attach to the bone close to where it originated. Fibers blend into other muscles close by and attach into the subcutaneous tissue and skin. The inner fibers of orbicularis oculi lie on the eyelids themselves, attaching to the medial palpebral ligament, which itself attaches to the bone just to the side of the bridge of the nose. You can feel the attachment of this ligament quite easily, just beyond the inner angle of the eye, the inner canthus. The Oxford English Dictionary, rather boringly, says that *canthus* is Latin for "the inner angle of the eye." But my old school Latin dictionary says something different: It says it means "the tire of a wheel" (and that

would have been an iron tire, holding the wheel together, not the rubber ones we have now, which are for shock absorption). In ancient Greek, *kanthos* seems to carry both meanings: the tire or rim of a wheel, and the corner of the eye. The association with "wheel" makes me wonder if this is something to do with orbicularis oculi itself, and its circling fibers. Digging deeper, another meaning for *canthus* is "corner" or the "lip of a jug" (and this is why "decant" is "pouring out"). This seems very apt, as tears fall from that inner angle of the eye.

The muscle name is somewhat simpler: a seventeenth-century invention, in Latin, drawing on *orbis*, a "circle" or "disc," and *oculus*, "the eye." Encircling the eye, orbicularis oculi closes the eye, bringing the eyelids together. The palpebral part closes the lids gently—a conscious blink, or as a reflex to protect the eye. Recruiting the wider orbital fibers scrunches the eye tightly closed, wrinkling the skin around it. Did you just blink?

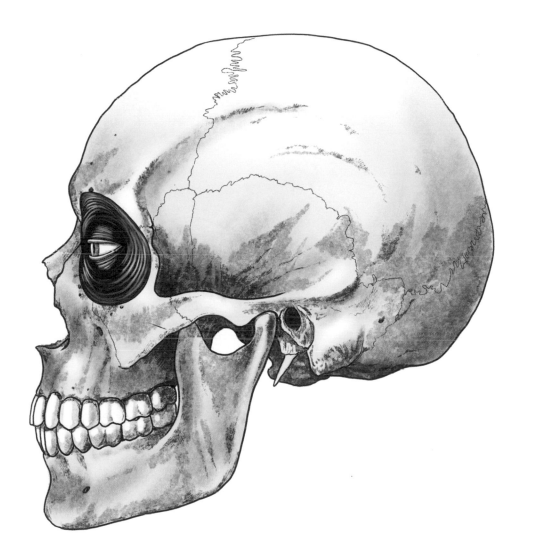

PAMPINIFORM PLEXUS

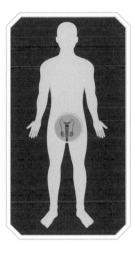

Somehow this sounds elaborate and extravagant, like the piled-up hair of Madame de Pompadour: a pimped-up, pumped-up pile of anatomy, a pampered profusion of something.

Let's deal with "plexus" first. A noun derived from the Latin verb *plectere*, "to plait," itself related to the ancient Greek *plekein*, meaning the same thing. Anatomical plexuses are intertwined strands—plaits—of nerves or, in this case, veins. As for the more flamboyant-sounding "pampiniform," this is a scientific Latin term that emerged in the seventeenth century, based on the classical Latin *pampinus*—meaning "vine shoot" or "vine tendril."

The pampiniform plexus is a twisting plexus of veins drawing blood from the testis and trailing up around the vas (or ductus) deferens (from the Latin *deferre*, to carry away). Eventually that convoluted confluence resolves itself into a testicular vein, which takes the blood up, up, with the right testicular vein emptying into the inferior vena cava, and the left into the renal vein.

Why such a tortuous plexus? Because, the tradition goes, the cool venous blood looping around the testicular artery brings down the temperature of the blood heading to the testis. And the testes hang around outside because they like to be cool; sperm only form properly in low temperatures. It sounds reasonable, until you consider that there's also a pampiniform plexus of veins draining the ovary—which is sensibly inside the female body (and perfectly happy making eggs at body temperature). I think the form of the pampiniform plexus is far more likely to be a quirk of development than an evolutionary adaptation to ensure the supply of cooler blood to the testis, even if the testes have needed to migrate outside the body to make sperm effectively in warm-blooded animals.

New balls please!

F ood is delicious, nutritious, packed with things that get your taste buds tingling and your digestive juices flowing; food sustains you and gives you the material to grow and repair your body, it provides you with energy and keeps essential cellular processes ticking along. But there's a fly in the ointment— or, rather, in your soup. When you put things from the outside world inside you, you may well be inviting in the enemy—in the form of all manner of malicious, pathogenic microbes. The body readies itself against the threat of invasion by parking its troops around those vulnerable areas where the outside comes very close to the inside of your body. Immune cells, including lymphocytes, lie in wait in barracks and fortresses known as lymphoid tissue. Lymph itself is tissue fluid that lies around cells, outside blood vessels. It's a Latin word, meaning "clear water," somehow related to the Greek word *nymphē*, "a water goddess." Your bone marrow and thymus are the training grounds

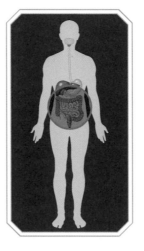

of lymphocytes, which then go and live in various fortlets known as lymph nodes (from the Latin *nodus*, for "knot"). Lymph flows into and out of lymph nodes via thin lymphatic vessels. As it flows through a node, lymphocytes watch the fluid very carefully for any sign of invaders.

Many lymph nodes are bean-size nodules, dotted around the body in superficial tissue, close to veins. Other lymph nodes guard the exposed tissues of the gut tube. The tonsils are effectively lymph nodes organized around the throat, in the walls of the pharynx and in the back of the tongue. And down in your small intestine, there are lymph nodes splatted into the walls of the gut. You don't want unfriendly bacteria colonizing your interior, and Peyer's patches are your guard against that. They're named after Swiss anatomist Johann Conrad Peyer, who first described them in 1677. There are about a hundred of them in your intestine when you're twenty years old, and somewhat fewer after that.

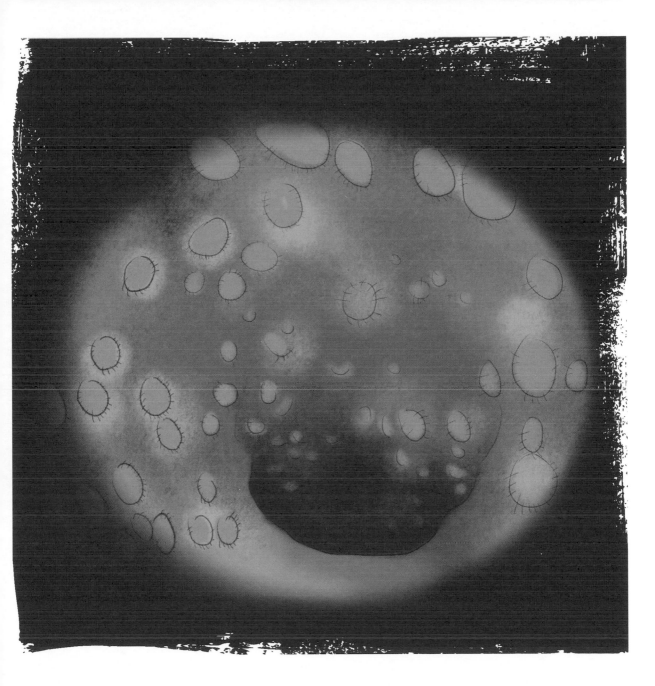

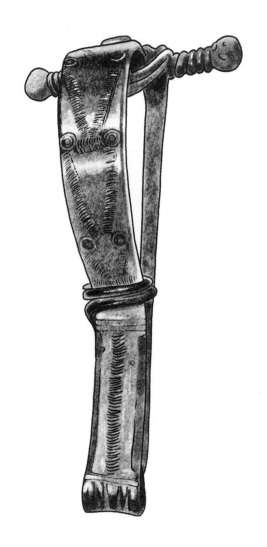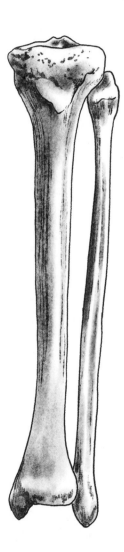

FIBULA

The bones of the leg have lyrical names. First, there's the weight-bearing shin bone, or tibia. *Tibia* means "flute" in Latin—apparently named by Aulus Cornelius Celsus in the first century, though who knows? He's the first to have written it down but I strongly suspect he didn't make up the name, as the use of animal tibiae as bone flutes goes way back. Sheep and goat tibiae were used in this way in ancient Greece, and across Europe, through to the Middle Ages—and are still used by some hunter-gatherer communities in Africa. Other bones were used, too—the oldest known bone flute, found in Geissenklösterle Cave in what's now Germany, was fashioned from a swan radius, forty-two thousand years ago. Early Chinese flutes were made from crane-wing bones. All these bones are perfect as wind instruments; the bird bones are hollow and ready to go, once the ends are broken off. A chunkier sheep tibia would need the marrow clearing out, but once that's done you have a hollow cylinder of bone, ready for the holes to be drilled.

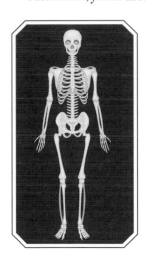

Meanwhile, *fibula* means "brooch," from the Latin *figere*, "to fix." (*Fix* and *figere* are both descendants of the same original Proto-Indo-European word.) The equivalent word in Greek was *peronē*—"pin"—which leaves its echo in alternative names for the long and short muscles that arise from the fibula: Fibularis longus and brevis are sometimes called peroneus longus and brevis. The peroneal nerve and artery are also known as the fibular nerve and artery. The fibula is like a brooch pin, lying alongside the tibia. But I wonder if there's something more to this name, too—that rather than just looking like a brooch pin, fibulae could have been used to fix clothing. A pointed chicken fibula would make a perfect ready-made pin.

Y ou may think that you don't have a tail. In fact, the definition of an ape (and you are one, sorry) is that it is a tailless primate. But actually you do have a very short little tail, and a tail bone. Three to five tiny, fused vertebrae finish off your vertebral column at the bottom—or rather, *in* the bottom. You can't wag it, or hang from trees by it like a spider monkey (more's the pity), but it's an essential piece of anatomy, not just a vestige of your former glory days when your ancestors did possess tails. Without a coccyx, you see, you would have nothing to attach your pelvic floor muscles to, at the back. Without that attachment, your pelvic viscera would start to fall out, and nobody wants that.

"Coccyx" is a funny word. It comes from the Greek *kokkyx*, meaning "cuckoo." Viewed from the side, it does look a bit like a bird's beak, I suppose. And that means you have two avian beaks in your body—one in the shoulder and another in the butt. Separated from the sacrum above, the coccyx has two little horns projecting upward—miniature versions of the superior articular

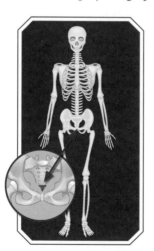

processes that most vertebrae possess, to articulate with the vertebra above. These *cornua* are simply named from the Latin for "horns," a name that also applies to the upper corners of the uterus, where the oviducts attach; to the horn-like projections of the lateral ventricles in the brain; to processes sticking up and down on the thyroid cartilage; and to the extensions of gray matter in a cross section of the spinal cord. *Cornu* comes from the Proto-Indo-European word **ker-*, meaning "horn," and actually "horn" comes from the same root, too; the Germanic languages seem to have lost the hard "k" and ended up with a softer "h."

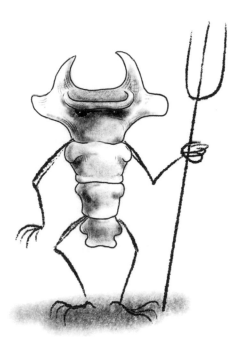

PILLARS OF THE FAUCES

At the back of your mouth, a double set of vertical ridges rise up on each side, from the base of your tongue, arching up to your palate. Nestled between these pillars are your palatine tonsils (often just called the tonsils, though there are others—pharyngeal, up behind your nasal cavities, and lingual, in the back of the tongue itself—all of them massed lymphoid tissue, helping to guard this opening into the interior of your body). The pillars are ridges lined with mucosa—overlying muscles that contract as you swallow, helping to push the lump of chewed food back, into the pharynx. "Pillars" is self-explanatory, but *fauces* is a Latin word, with a seemingly straightforward translation: "throat." It always seems to be the plural, *fauces*, that's used in this way. The singular, *faux*, could also be used to describe a chasm or gorge in a

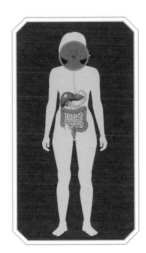

landscape, and sometimes even apparently inverse topography: an isthmus or "neck" of land. The French word *faux* is almost certainly nothing to do with the Latin for "throat"—it comes instead from *falsus*, which for once is more recognizable in non-Romance English: false. Though perhaps there is something false after all about the pillars of the fauces—they are not freestanding columns, but pilasters, firmly stuck to the walls of the throat.

Another avian delight, this piece of anatomy is named after the foot of a goose, in Latin. It's the point of insertion on the medial surface of the tibia for three long tendons: those of sartorius, gracilis, and semitendinosus muscles. As they approach their destination, the tendons merge, their fibers enmeshing with each other, and then insert into the bone. In the dissection room, you can dissect and separate out these converging tendons, and then the insertion takes on the appearance of a webbed foot. Sort of.

The thigh contains three groups of muscles: an extensor group in the front, an adductor group on the inside of the thigh, and a flexor group (the hamstrings) on the back. The pes anserinus involves one muscle from each of these compartments: sartorius from the front, crossing diagonally over the quads; gracilis running down the medial thigh; and semitendinosus from the hamstrings. The sartorius muscle flexes the hip joint, laterally rotates the thigh, and flexes

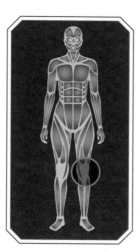

the knee. If you want to know what that looks like, sit cross-legged on the floor—and this is apparently how tailors traditionally used to sit to do their sewing; *sartor* is Latin for "patcher" or "mender." Gracilis is, as its name suggests, gracile—*gracilis* meaning "slender" in Latin. Semitendinosus is fleshy above and narrows into a slim tendon below; lots of muscles are half-tendinous in this way, but this is the only one named for that particular attribute. These three muscles together are sometimes called the "guy ropes" of the thigh.

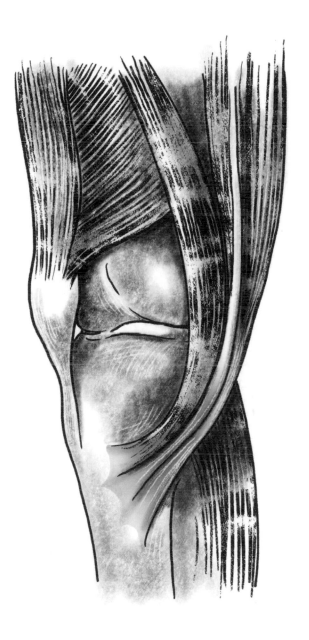

MONS PUBIS

Mounded up on top of the joint between the pubic bones is a pad of fat, forming the mons pubis: the "pubic mountain." Sometimes this landmark is more lyrically called the mons veneris: the "Mountain of Venus." The fat pad is there in both sexes, but more prominent in females. Naturally forested in pubic hair, the shape of the mons pubis may be laid bare by Brazilian waxing.

Pronouncements about the function of anatomical features around the pubis and perineum tend to focus on potential roles during sex. The mons pubis is said to be there to provide a cushion during sex, but this of course assumes that a missionary position, with the man on top, has been so prevalent during human evolution as to create a selection pressure for this fat pad to develop in females. Pubic hair has, similarly, been suggested to reduce friction and the risk of chafing.

Pubic hair develops as a secondary sexual characteristic, which may link it to sexual behavior—perhaps. But then what about armpit hair? Do we assume that's all about the act of sex as well? Are armpits ever at risk of chafing during intercourse?

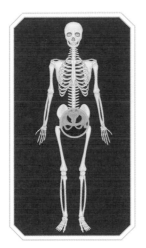

It seems pertinent that no other primates have this forest of hair around their pubic region (though they are generally more hairy than us all over)—and in fact they are mostly *less hairy* in that area than we are, as adults. The forest of pubic hair that develops on the mons pubis in humans could be there to signal sexual maturity, and perhaps to trap musky pheromones as well. These seem more likely explanations than chafing-protection, but we may never truly get to the bottom of this conundrum. The etymology of *pubis* is easier to bottom out; it means all of "pubic hair," "groin," "adult," or "full-grown" in Latin.

MASSETER & TEMPORALIS

Ruminate a little on the meanings of anatomical terms and sooner or later you come to chew on this one. The muscles of mastication are the ones that move your lower jaw and bring your teeth into grinding opposition. (Just the opening syllable of the word is enough to elicit a giggle from nervous first-year medical students, who are just beginning to realize that they will have to name everything, absolutely everything, and will not be able to hide behind euphemisms or slang any more). "Masticate" comes from Latin, *masticare*, which comes from Greek *mastikan*, based on *mastax*, "jaw." (As opposed to "masturbate," which is suggested to have come from the Latin words *manus* and *turbare*—to "hand-disturb.")

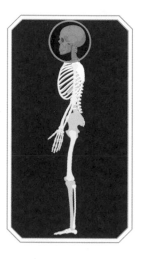

There are two muscles of mastication on the inside of the jaw: the pterygoid muscles, which spring from the pterygoid process of the sphenoid bone in the center of the skull. And then there are two muscles on the outside: the masseter, which attaches from the zygomatic arch to the angle of the mandible, and the temporalis muscle, which attaches from the side of the skull to the lever-like coronoid process of the mandible. Both these muscles pull up on the jawbone and close the teeth together. "Masseter" is straightforward: a name from ancient Greek, *masseter*—"chewer"—via Latin. And that word in Greek must surely be related to *masso*, "to knead." The muscle temporalis takes its name from the bone of the skull that it arises from, the temporal, from Latin *tempora*. This is a plural, with the singular being *tempus*. This seems to come from Proto-Indo-European **temp-*, "to stretch"—in the sense that the temples are the sides of the forehead where the skin is thinly stretched, above the cheekbone. But there's another suggestion, tied up with the synonymous meaning of *tempus*—"time." It is the hair over our temples that first shows the signs of time passing, turning white with age. Place your fingertips on your temples and chew—you can feel the fibers of temporalis muscle contracting, under the skin. Each contraction takes time, bringing you a fraction of a second closer to the end of your life . . .

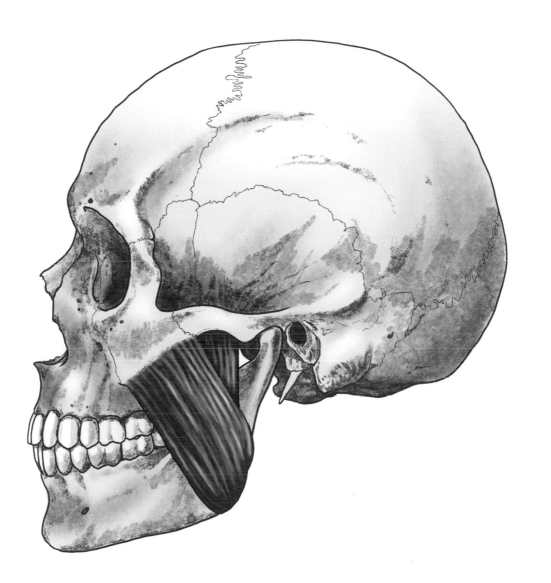

Yet another bird word lurks inside the skull: *Crista galli* is Latin for "a cock's comb." It's part of the ethmoid bone, which sits above your nasal cavities, between your eyeballs and under the frontal lobe of your brain. This curious bone includes sets of air cells on each side: The ethmoid sinuses, which, just like the maxillary sinus, are lined with respiratory (ciliated, mucus-manufacturing) epithelium. And once again, being paranasal air sinuses, they open into the nasal cavity; some into the superior, others into the middle meatus. (*Meatus* is Latin for "pathway"—from *meare*, "to pass"—and is used for these channels of the nasal cavity, defined by the conchae projecting in from the side wall, and also for the canal of the ear, and the urethral opening on the tip of the penis.)

At the top of the ethmoid bone is the cribriform plate, a horizontal sheet of bone peppered with holes. It forms the roof of the nasal cavities, which are very narrow in these uppermost reaches. The holes are for olfactory nerves—the nerves that carry signals from

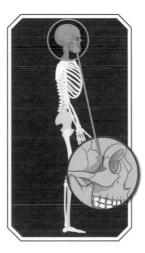

olfactory receptors in the nasal cavity. That word comes from the Latin *olfacere*, "to smell," which itself comes from *oleo*, "to emit an odor" (which could be fragrant or stinking). This is very close to the word *oleum*, a Latin word related to the Greek *elaion*, and both meaning the same thing: oil; usually olive oil. Today, we might think of this as a purely culinary substance, but in ancient Greek culture it was much more used for lamps, soap, and for rubbing on the bodies of athletes—and the dead. It was fundamental to the economy; Homer called it "liquid gold." The passages for the olfactory nerves, which allow us to smell olive oil and many other scents, provide a name for this part of the ethmoid, as well as for the whole bone. *Cribriform* means "sieve-like" in Latin; *ethmoid* means "sieve-like" in Greek. The cribriform plate of ethmoid is the sieve-like plate of the sieve-like bone.

Mounted atop the cribriform plate is the vertical crista galli, a projection of bone providing attachment for a thick sheet of connective tissue—part of the dura mater—lying between the two hemispheres of the brain and known as the falx cerebri.

VENTRICLES

The brain is not a solid lump of nervous tissue; it is a hollow organ. The spaces within it, which started out life in utero as a simple canal in the center of the neural tube that is the first semblance of the central nervous system, are pulled into curious shapes as the brain grows and the cerebral hemispheres bud and expand. There are two lateral ventricles, which are connected to the central third ventricle, which is itself connected via the cerebral aqueduct to the fourth ventricle—which leads into the narrow, central canal of the spinal cord. Each ventricle contains a skein of capillaries covered with a web-like layer of pia mater, and this choroid plexus makes the cerebrospinal fluid which fills the ventricles and then flows out around the brain and spinal cord. The word "choroid" is used to describe connective tissue containing lots of blood vessels; it's also used for the highly vascular layer in the coat of the eyeball. It

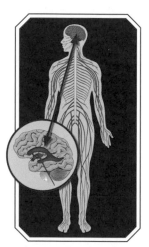

seems to have come from a Latinized corruption of the Greek word *chorion*—originally describing the outer membrane around a fetus.

"Ventricle" comes from Latin, *ventriculus*—the diminutive of *venter*, "belly." So it literally means "little belly"—and was once used as a term for the stomach. It's still used to describe the lower chambers of the heart and the little pockets in your larynx, just above the vocal cords. And how strange, to think that in your brain there are four "little bellies."

BRACHIAL PLEXUS

When your muscles and the nerves supplying them start to grow in the embryo, it's all very straightforward, neat, and tidy to begin with. At just five weeks old (after conception), a series of bead-like lumps would have been visible, all the way down your back, on either side of the ridge formed by the neural tube. These lumps were somites—body segments, from the Greek *soma*, "body," and the suffix *-ites*, "belonging to" (the same root as the almost always inflammatory "-itis"). Each lump contains three parts, which begin to separate away from each other and head for their ultimate destinations: One part will form the lower layer of your skin, the dermis; another will form bone; and another will form muscle. The name for this muscle-forming part of the somite is the "myotome." This comes from Greek—*muo-*, for "muscle" or "mouse," and *tomos*, "cut" (the same suffix that gives us "anatomy"—the science that cuts apart). A single spinal nerve grows out from the spinal cord to innervate each myotome block. If only it stayed that simple. It does in the chest, where the muscles between one rib and the next come from one original myotome and are supplied by a branch of the spinal nerve that simply runs around the body wall, just under each rib.

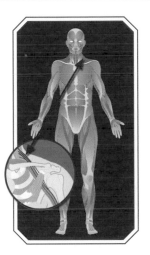

But we're not worms, and it all gets very complicated when limbs begin to form in the embryo. They appear as small buds, then grow out, gradually developing elbows and knees, wrists and ankles, and flat plates of tissue at the ends which will eventually form hands and feet, with separated fingers and toes. Inside the limb, loose embryonic connective tissue condenses and forms cartilage models of the bones that will develop in due course. And bits of myotome migrate into the limbs, splitting and fusing with each other to form individual muscles. It's rare that any limb muscle derives from just one myotome; most are formed by contributions from two or three. Their nerves come with them, and rather than being supplied by separate branches from two or three spinal nerves, the roots of these nerves come together to form a skein, where fibers are swapped. The peripheral nerves that emerge from the skein contain contributions from spinal nerves matching the myotomes in the muscle they're heading to.

The limb plexus for the arm is brachial, from the Latin *brachium*, from the Greek *brachion*, "arm." Just as we can trace the spinal-nerve roots of each peripheral nerve, we can chase down the origin of these terms. The word *plexus* is Latin for "plaited." The Greek, *plektos*, is similar, and goes all the way back to Proto-Indo-European, and the root **plek-*, "to plait."

CALYX

All the time, your kidneys are busy making urine: filtering blood, getting rid of urea, and carefully getting the balance of water and sodium, and all manner of other ions, just right. The stats are staggering. The 1.3 gallons (5 liters) or so of blood contained in your circulation pass through the kidneys around 300 times each day. That means the kidneys effectively filter some 400 gallons (1,500 liters) of blood daily. Around 40 gallons (150 liters) passes out of the glomerular capillaries and into Bowman's capsule. *That's too much!* I hear you cry. But as that raw urine passes along the renal tubule, including its loop of Henle, 99 percent of the fluid is reabsorbed into the blood, leaving you with a much more manageable 50 ounces or so (about a liter and a half) of "finished" urine to expel from your body, and the bladder can hold up to 16 ounces (half a liter) at a time. The renal tubules empty into collecting ducts in the medullary pyramids of the kidney, and the ducts join together until there are around twenty left to open on the apex of

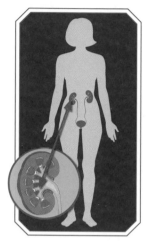

the pyramid—which is called the renal papilla (the obsession with teats reappears!). From there, the urine enters a cuplike space called a minor calyx, the stem of which joins with those of other minor calyces to form a major calyx. There are around ten minor calyces, joining to make two or three major calyces, in each kidney. The major calyces then coalesce to form the renal pelvis, which drains into the ureter, which carries the urine down to the bladder.

Calyx is a Latin word, from the Greek *kalyx*, and originally meant "shell" or "husk," from the Greek *kalyptein*, "to cover." But in medieval Latin, it became conflated with the word *calix*, which means "goblet." Certainly, the way it is used for these structures in the kidney is much more to do with that idea of a "cup" rather than a "covering." In botany, the original sense is kept—the calyx forming the outer covering of a flower bud. And "calyx" is also an old word for the follicles in the ovary, where the ova lie, and this creates a beautiful word-picture of the follicle opening like a flower at ovulation to release the egg.

ZYGOMA

T his is my second-favorite bone of the skull after the sphenoid. And the temporal. Okay, third-favorite. The sphenoid and the temporal bones have awesomely intricate shapes—and fascinating etymologies. "Temporal" has been dealt with (see page 98). "Sphenoid" is Greek—with the *-oid* bit meaning "-like" or "form," as usual. And *sphen* means "wedge." So it's the wedge-shaped bone (though I think it looks more like a bat) and it really is jammed into the center of the skull. You can see bits of it from so many angles: at the back of the eye socket, on the temple (just in front of the temporal bone), and on the base of the skull. The zygoma is less exciting in terms of its shape—though quite elegant, being triangular with extended points. It lies to the side in your cheek; sometimes it's called the malar, from *mala*—Latin for "cheek." One

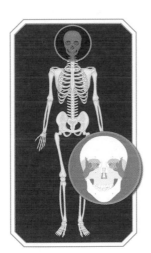

of the points of the zygoma projects backward to connect with a slender process of the temporal bone, forming the zygomatic arch, more familiar as the cheekbone. Its other points connect to the maxilla (upper jaw) and the frontal bone (which underlies your forehead). It's almost *harnessing* those bones of the face back to the side of the skull—and that's exactly what its name means: "zygoma," *zygon* in Greek, is "yoke."

We're back inside the skull—a skull in which the brain has been removed, but the dura mater left in place. And now we can see the great curving sheet of dura that stretches forward—running in the longitudinal fissure that exists between the two hemispheres of the brain—and attaches to the crista galli at the front. It's like an arcing blade, and its name is very fitting: *falx* is Latin for "scythe" or "sickle." Anatomists argue about why it's

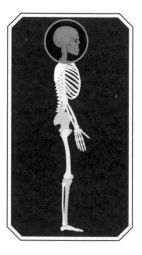

there. It might help to stop the brain swiveling inside the skull. The double-dural membrane that forms the falx cerebri encloses two large veins, the superior sagittal sinus above (right up against the skull) and the inferior sagittal sinus below, in the free edge of the falx. These veins are described as venous "sinuses"—not to be confused with the sinuses of the nose. As they lie within dura, their walls are rigid—unlike the floppy and collapsible walls of most veins.

The two sagittal sinuses connect with other sinuses, draining blood from the brain and cranium. The inferior sagittal sinus flows back into a straight sinus (called—uncharacteristically simply—the straight sinus), which joins the superior sagittal sinus right at the back of the skull, where the occipital and transverse sinuses also meet. This junction is called the confluence of the sinuses, or even more poetically, the torcular herophili—literally, the "wine press of Herophilus," a strangely macabre name for this tangle of channels, making us think of the wine-dark venous blood being squeezed out of the cranium, eventually flowing out through the internal jugular vein, exiting through the base of the skull.

But it seems that layers of translation, from the original Greek, have added a certain twist to the tale. The word Herophilus himself used for the junction of these membrane-bound vessels—as recorded by Galen—was *lenos*, "wine trough." The meaning seems to have changed from "trough" to "press" as the anatomical term passed into Arabic as *al ma sara*, a word that initially meant "vat" but transmuted over time to mean "an olive press." When that word was translated into Latin in the twelfth century, it became *torcular*, itself based on the verb *torquo*, "to twist." Herophilus, too, must be turning in his grave.

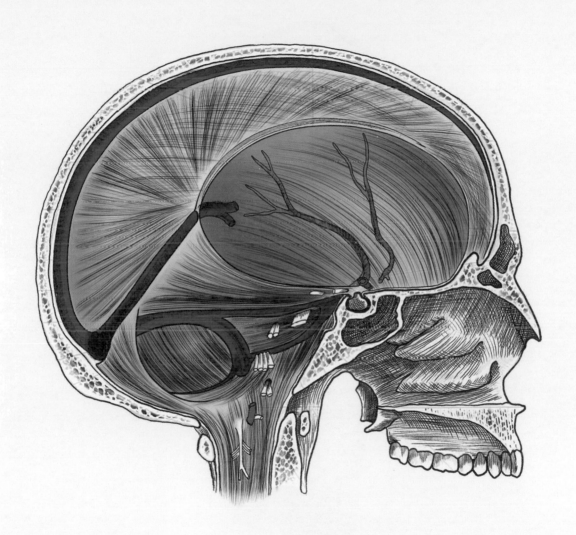

AMPULLA OF VATER

Sounding like an archvillain in a space opera, the ampulla of Vater is much smaller than its epic-sounding name suggests. It represents another sort of confluence (see the preceding entry), this time of two ducts carrying digestive juices: the common bile duct and the pancreatic duct. The ampulla is just 2 to 10 millimeters long, and it opens on the apex of the major duodenal papilla, halfway down the descending part of the twelve-fingers-long duodenum. At least, that's what it does in around two thirds of people. The other third don't have an ampulla of Vater: The bile duct and pancreatic duct open separately on the papilla, or even—more rarely—at two completely separate places in the duodenum.

The ampulla is named after Abraham Vater, an eighteenth-century German anatomist who studied at the famous University of Leipzig before eventually becoming a prof at the University of Wittenberg. He described the hepatopancreatic ampulla in his *Dissertatio anatomica*

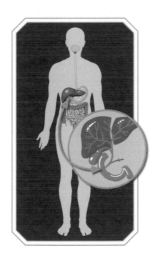

quo novum bilis dicetilicum circa orifucum ductus choledochi ut et valvulosam colli vesicæ felleæ constructionem ad disceptandum proponit. I don't think he was known for his concision. But the Latin word *ampulla* was a good choice for a short, conical channel. It means "flask" or "bottle," and also gives us the word "ampoule," which is often used to describe the small, sealed bottles that vaccines are stored in. A Roman ampulla was a small bottle—often glass, with two handles—that you could have used to carry your oil with you to the baths.

T his is a fantastic geometric term that I'm sure should be more widely used outside of anatomy. It comes from the Latin, *triquetrus*, meaning "three-cornered." The *-quet-* bit may come from a Proto-Indo-European root, *$\hat{k}\bar{e}(i)$*, which also gives rise to "whet"—to sharpen. "Triquetral" is a sharp and snappy alternative to the more quotidian "triangular." In anatomy, it's used regularly to refer to the small, three-cornered bone in the carpus, on which the pisiform bone sits.

"Triquetral" can also be used for little triangular bones that are occasionally found within the joints, or sutures, of the flat bones of the skull, although these are more often called Wormian bones—named after the seventeenth-century Danish physician Olaus Worm. He was a bit of a hero, choosing to stay in Copenhagen to minister to plague victims—rather than fleeing the city like so many others. Eventually, he would die of the plague himself. As well as making contributions to the sciences of embryology and anatomy, he also proved that unicorns did not exist—or at least that the horns widely supposed

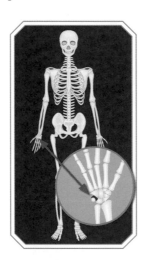

to have come from these mythical beasts were actually from narwhals (something I suspect many sailors already knew, but still enjoyed spinning a tall tale).

"Triquetra" is also the name for a knot of three interlaced arcs, creating a three-cornered shape—sometimes known as a triple knot—often seen in the insular art that developed in Britain and Ireland in the first millennium of the Common Era. Those knot patterns had their roots in Roman art—including the plaited designs popular in mosaics—and must surely have been based on textiles: on decorative knotwork and, later, lace-making.

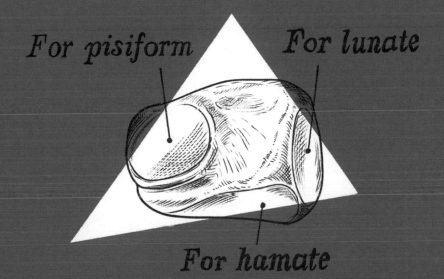

For pisiform

For lunate

For hamate

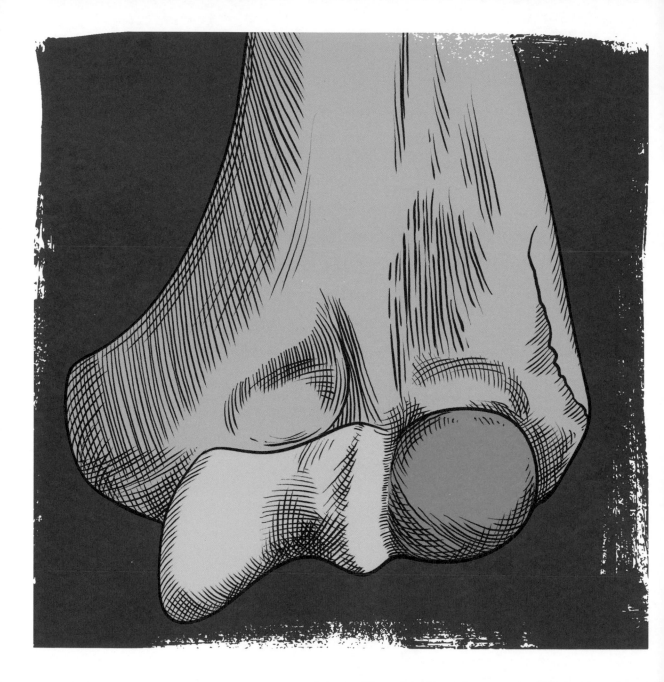

TROCHLEA

his is where the ulna articulates with the humerus, at the elbow. The distal humerus has a complex shape, with a small, round capitulum (Latin for "little head"), which articulates with the radius, and a spool-shaped trochlea, which forms a close-fitting joint with the upper part of the ulna, including the anterior surface of the olecranon, which together form the reciprocating trochlear notch.

Trochlea is a Latin word, based on the Greek *trochilia*, meaning "pulley." This part of the ulna really does look like a pulley—or a diabolo. The word pops up in other places in the body. The bottom end of the tibia, where it articulates with the uppermost bone of the foot, the talus (Latin for "ankle bone"), was sometimes referred to as a trochlea. And there's a muscle inside the eye socket that used to be called the trochlearis muscle. This muscle runs from the back of

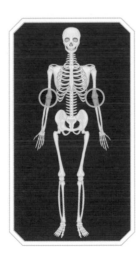

the eye socket, forward along the medial (inner) wall of the socket, then passes through a cartilaginous loop—a pulley—to change its angle and turn back on itself, running laterally to insert into the eyeball. It pulls the eyeball down and in—the direction your eyes swivel in when you walk downstairs. The new name for the muscle is the superior oblique, but its pulley is still memorialized in the nerve that supplies it: the trochlear nerve.

STYLOID PROCESS

Very few of the skulls that we used to teach the medical students at Bristol University still possessed their styloid processes. The skulls—all of them real—were treated with great respect, but every time they were moved in and out of cupboards, and in and out of cardboard boxes and large plastic trays, there was the chance that fragile pieces of bone would be broken. There were a select few that were only released from their bespoke wooden boxes on special occasions: the Exam Skulls—and these had perfectly preserved styloid processes.

The Latin word *stylus* is a variant spelling of *stilus*—which means a "sharpened stick," a "stake," or the pointed instrument used to write on wax tablets. It's another Latin word that owes a lot to Greek: *Stilos* in ancient Greek meant "pillar." There's a sense of something long and possibly cylindrical(ish) there, but the meaning becomes more *pointed* in Latin. Much later, a point used for etching copper plates would be called a stylus or a steel. Perhaps the instrument gave its name to the

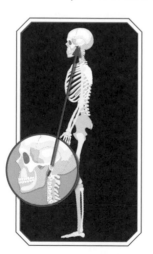

metal that was eventually used to make it, but the early ones, for writing on wax, were made of bronze.

The styloid process of the skull descends from the temporal bone, slim and pointed and just over three quarters of an inch (2 cm) long. Like so many bony projections in the body, it's the site of attachment of ligaments and muscles, and all of them take their name from it. There's the stylohyoid ligament, the stylomandibular ligament, and the styloglossus, stylohyoid, and stylopharyngeus muscles. With all those attachments, there really is a point to the styloid process.

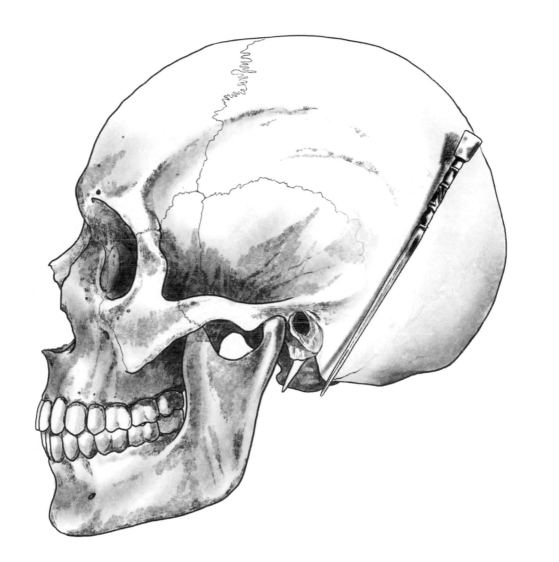

SCIATIC NERVE

T his is the nerve that emerges under piriformis muscle in the buttock and then runs down the back of your thigh, breaking into branches that supply the hamstrings, and then the muscles lower down, in the calf, all the way down into the foot and its four layers (yes, four layers—read it and weep) of muscles.

It has a lot of work to do, a lot of muscle fibers to innervate, so it's not surprising that this is the largest nerve in the whole body. It's three quarters of an inch (2 cm) wide and about a fifth of an inch (half a cm) thick. You might think of nerves as being thin little wires. The sciatic nerve is very definitely not that. Compression and irritation of the nerve—or of its lumbar nerve roots—leads to the characteristic shooting pain of sciatica down the back of the leg. It is not pleasant, as I can personally attest. It can take your breath away when you're standing up, giving a lecture on anatomy. But perhaps the origin of this nerve's name can help to take the pain away. It's a tortuous etymological path (if not quite as twisted as the torcular herophili).

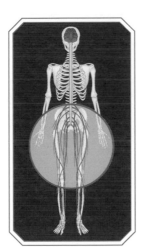

"Sciatic" comes from French, *sciatique*, from post-classical Latin *sciatica*; the name for the pain seems to have predated the name of the nerve itself. An earlier Latin term was *ischiadicus*, "hip pain," from the Greek *iskhiadikos*; *iskhion* means "hip joint." Part of the pelvic bone is called the ischium, from the same root. Whatever it goes on to do in the thigh, leg, foot—the name of the sciatic nerve goes right back to its anatomical origin, close to the hip joint.

Y ou can create the shape of the female reproductive system with your own body. It would make for a brilliant piece of fancy dress, and some day I will make this costume for myself. Bend your knees and stick your butt backward: Your legs represent the vagina, and your butt the cervix (Latin for "neck") of the uterus. Your trunk is the body of the uterus; tuck your head in, as it's just part of the fundus of the

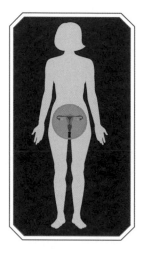

uterus. (*Fundus* means "bottom" in Latin. If the uterus is conceived of as a flask, with the opening being the cervix, then the fundus lies opposite this. In the body, though, the uterine neck lies lowermost and its bottom, its fundus, uppermost.) Now stretch out your arms, and if you can find a honeydew melon to hold in each hand, better still. Your arms are the oviducts, aka uterine tubes, aka Fallopian tubes, and the melons are the ovaries.

Vagina is a Latin word meaning "sheath" or "scabbard." The violence of sex is written into female anatomy in a disturbing way; the vagina is named for its role in coitus, with a term heavy with military connotations, and not

for its role in childbirth. *Uterus* is Latin for "womb" or "belly" and is also related to *uter*, Latin for an animal skin used as a bag. (And "womb" itself is a Germanic word, most closely related to Old Frisian *womme*, meaning "belly"). "Uterine tube" is self-explanatory. "Oviduct" means the path that the egg is drawn along. The Latin word *ductus* comes from *ducere*, "to draw along"; the patriarchal resonances in this verb are strong—it was used to mean "to lead," "to command," or "to take a wife." *Ovum* is Latin for "egg"—and both words come from the same Proto-Indo-European root, **owyo-/*oyyo*. *Ovarium* is simply post-classical Latin for the place containing the eggs, in the same way that an aviary, an *aviarum*, is the place where birds, *aves*, are kept. As for the eponymous name of the oviduct, that memorializes a sixteenth-century Italian anatomist, Gabriele Falloppio, who was once Professor of Anatomy at the famous University of Padua.

And with these organs that nurture the origin of every human life, we reach the end of our tour of anatomical oddities. The end brings us to the beginning.

for my
next
trick

ACKNOWLEDGMENTS

This little book draws on my own enduring fascination with digging into the history of words and finding unexpected gems of understanding and weirdness in there. I was further encouraged in this archaeology of anatomy by my friend and mentor, classicist and anatomist Dr. Jonathan Musgrave, when we worked together in the Anatomy Department at Bristol University. And I owe Jonathan enormous thanks for his painstakingly careful reading of the manuscript for this book, ensuring that I'd got my Greek grammar ironed out and my omicrons and omegas, epsilons and etas correctly translated. Any mistakes still in here are mine and mine alone. I must also thank the whole team at my original publisher, Simon & Schuster UK: my patient copy editor, Alex Newby, proofreader Jonathan Wadman, project editor Kaiya Shang, fabulous editor Holly Harris, brilliant designer Doug Kerr, and my fantastic literary agent, Luigi Bonomi.

INDEX

ABOUT THE AUTHOR

ALICE ROBERTS is an academic, author, and broadcaster, specializing in human anatomy, physiology, evolution, archaeology, and history. In 2001, Alice made her television debut on Channel 4's *Time Team*, and went on to write and present *The Incredible Human Journey*, *Origins of Us*, and *Ice Age Giants* on BBC2. She is also the presenter of the popular TV series *Digging for Britain*. Alice has been a Professor of Public Engagement in Science at the University of Birmingham since 2012.

alice-roberts.co.uk | theAliceRoberts | prof_alice_roberts

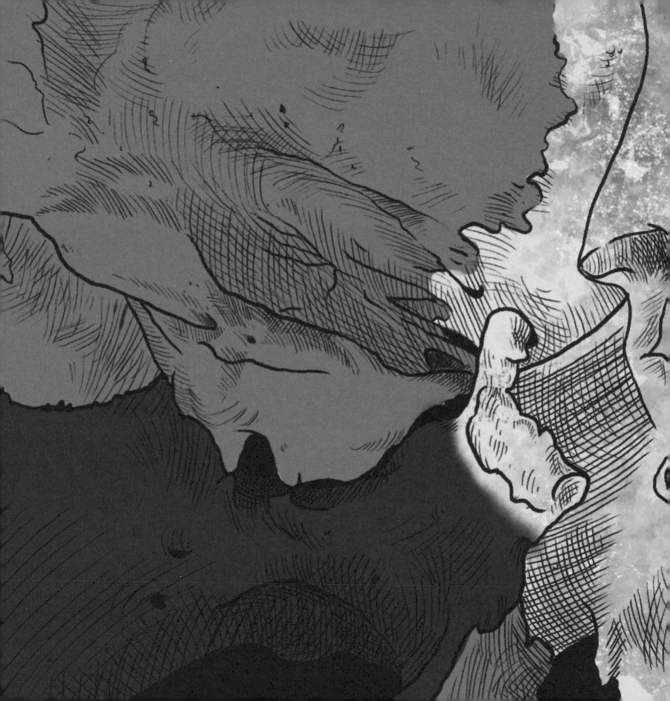